森林的故事

竹子篇

杨青 主编

中国林业出版社
China Forestry Publishing House

图书在版编目（CIP）数据

森林的故事.竹子篇:汉、英/杨青主编.--北京：
中国林业出版社,2022.12
ISBN 978-7-5219-2072-7

Ⅰ.①森… Ⅱ.①杨… Ⅲ.①森林—儿童读物—汉、
英②竹—儿童读物—汉、英 Ⅳ.① S7-49

中国国家版本馆 CIP 数据核字 (2023) 第 001000 号

森林的故事·竹子篇

数字资源信息

策划编辑：于界芬　吴　卉
责任编辑：吴　卉　于界芬　倪禾田
出版发行：中国林业出版社
（100009，北京市西城区刘海胡同7号，电话83143552）
电子邮箱：books@theways.cn
网址：https://www.cfph.net
印刷：河北京平诚乾印刷有限公司
版次：2022年12月第1版
印次：2022年12月第1次印刷
开本：787mm×1092mm 1/16
印张：17
字数：147千字
定价：88.00元（中英文版共2册）

丛书主编：曹福亮

主　　编：杨　青
副 主 编：曹　林　王　涛　贾文婷

编委会成员及分工
统筹策划：周吉林　杨　青
故事创作：杨　青　包心悦　李杨楠　曹曼曼
插图创作：耿植荣　贾文婷　赵亚洲　杨　青　宋昱萱
英文翻译：王　涛　曹　林
科学顾问：丁雨龙　张春霞　杨　建
艺术顾问：卫　欣

序

竹子是一种神奇的植物。1982年,我从南京林业大学毕业留校在竹类研究所工作,研究的对象就是竹子。那时我经常和周芳纯先生一起到江苏宜兴、浙江莫干山、江西宜丰、广西桂林等竹子产区做号竹、土壤取样、竹资源调查、产量估算等科研工作,往往一待就是一两个月,就这样和竹子打了4年多的交道。近距离地接触,使我对竹子有了较为深刻的认识。随着相关知识的不断积累与丰富,我更加意识到竹子的经济价值、生态价值、文化价值在植物界举足轻重。源于这种对"竹子"的情感,结合自身工作、研究的经历,本着苦中作乐的奋斗精神和严谨端正的科学态度,我们决定为少年朋友们写一本关于竹子的科普书,让大家通过这本书去认识竹子。因为,通过老师们深入浅出的科普,你们能够了解竹子,提高自己的科学素养。这有益于你们的成长,有益我们民族的未来,有益地球的未来。

竹子是个大家族,全世界有1400多个品种,分布在热带、亚热带、温带。中国是竹子的故乡,有600多种。考古发现,竹子在我们祖先的生活中有着重要作用,搭桥造屋,做家具,做农具,竹笋还能做成美味佳肴。尤其是在商周时期,聪明的祖先用竹子做成竹简，代替了甲骨，成为记录文字的重要工具。

在地球日益变暖、环境问题突出的今天，竹子算得上是人类的好伙伴。竹子的水土保持能力、吸收二氧化碳能力、固碳能力、放氧能力都远高于其他植物。竹子种植后，3到5年即可间伐。竹子作为工业用材，经济效益很好。竹子有亭亭玉立的身材，纤细玲珑的翠叶，挺拔虚心的枝干，不择土壤生长的特性，和中华民族的很多美德相似，所以，竹子也成为中华传统文化歌颂的对象。

用植物学的视角看竹子，我们会发现竹子有着许多耐人琢磨

的特点，令人着迷：竹子到底是草还是树？竹子为什么长得特别快？竹子为何长高不长粗？还有，为什么一株竹子日后就能长成一片竹林？竹子为什么终生只开一次花，开花后还会死？类似这样的问题在这本《森林的故事·竹子篇》里，大家都能找到答案。

  少年朋友们，竹子事业大有可为！1997年，国际竹藤组织在中国北京成立，这是第一个把总部设在我国的政府间国际组织，目前已经有了48个成员国。它既标志着我国在竹藤研究方面的重要影响力，也说明了从事竹藤文化研究的广阔前景。我希望，你们从这本书中认识竹子，喜欢竹子，继而有着更多的探索植物奥秘的想法和行动，用你们的努力让大地更绿，山河更美。

曹福亮
2022年10月

前言

亲爱的少年朋友们，这是一本关于竹子的科普读物，使用老少对话的方式，采取中英对照的文字，辅以手绘插图和音频朗读，介绍竹子的相关知识，风格清新活泼，表达妙趣横生，读起来轻松愉快，思考后脑洞大开，举一反三。阅读这本书，能给你带来意想不到的收获。

竹子是我们熟悉的一种植物。植物学家说它是高度木质化的草本植物，环保人士说它是和地球环境最友好的植物，文人墨客说它是坚贞、刚毅、挺拔、虚心、有节等美德的化身，还给了它"篁""筠""抱节君"等诸多美称。

本书分成六个部分。

读"观竹——胸中有成竹"，我们可以明白竹子为什么能有好身材，懂得为什么雨后春笋节节高，清楚竹鞭在地下的行走方式，知道竹子的很多生长奥秘。

读"赏竹——青青四季同"，我们就像走进竹子博览会，见识竹中之"最"，延绵似藤的攀缘竹，身洒泪珠的斑竹，嵌金镶玉的竹，都历历在目。

读"种竹——无竹令人俗"，我们可以了解一些竹子栽培经营的科学知识，了解克隆技术在竹苗培养中的应用，甚至可以学习一下怎样挖笋。

读"用竹——不可居无竹"，我们知道竹子不仅是制作建材、家具、药品、乐器的原料，而且通过现代科技让竹子变竹炭净化空气、把竹纤维做成纺织品。研究竹子，已经受到多国科学家们的重视。

读"爱竹——情寄幽篁间"，我们可以了解竹子在中国文化中的特殊地位，它和"菊""梅""兰"并称为"四君子"，却不止在书画中出现，它还是文化的传承者，民俗的承载物。

读"护竹——大熊猫与竹"，我们了解大熊猫和竹子的故事，知道大熊猫在进化中选择竹子作为食物是很不容易的，我们应该努力保护竹子，为大熊猫，也为自己创造美好生活。

少年朋友们，我们都读过郑燮的诗，"咬定青山不放松，立根原在破岩中，千磨万击还坚劲，任尔东西南北风。"竹子的坚忍执着曾引起我们的赞美尊重。现在，让我们开心地读读这本《森林的故事·竹子篇》，使有关竹子的科学知识在我们的脑海中安营扎寨，成为我们成长的助力。如果通过本书的阅读，有少年朋友爱上竹，将来成为从事和竹子有关的某个方面的专业人士，那是我们共同的幸运。

2022 年 10 月

目 录

序　Ⅳ
前言　Ⅶ

观竹——胸中有成竹　1
1. 非草非木的竹子　2
2. 竹子开花也结实　5
3. 永葆身材的竹子　8
4. 任意行走的竹鞭　11
5. 雨后春笋节节高　14
6. 神奇的竹笋吐水　17
7. 竹笋竹子是同辈　20

赏竹——青青四季同　23
8. 竹中之最　24
9. 竹门三家族　27
10. 延绵似藤的攀缘竹　31
11. 腰肥肚圆的佛肚竹　35

12. 嵌金镶玉的竹子　　38

13. 母慈子孝的孝顺竹　41

14. 身洒千滴泪的斑竹　44

种竹——无竹令人俗　　47

15. 竹子种植讲科学　48

16. 科学克隆小竹苗　51

17. 竹笋采挖有技巧　54

18. 毛竹标号建档案　58

19. 竹子钩梢抗雪灾　61

用竹——不可居无竹　　63

20. 一竿为舟的竹　64

21. 置景园林的竹　68

22. 钢筋铁骨的竹　71

23. 变身家用的竹　74

24. 变形柔丝的竹　77

25. 可制良药的竹　80

26. 净化空气的竹　83

27. 变幻妙音的竹　86

爱竹——情寄幽篁间　89

28. 国画中的竹子　90
29. 古诗中的竹子　93
30. 物以载文的竹简　96
31. 爆竹的前世今生　98

护竹——大熊猫与竹　101

32. 选竹为食的大熊猫　102
33. 食竹需要强大肠胃　105
34. 握竹特化长出六指　108
35. 竹子开花预示饥荒　111
36. 护竹创造美好未来　114

参考文献　118
后记　119

观 竹
胸中有成竹

森林的故事·竹子篇
The Story of the Forest · Bamboo Chapter

非草非木的竹子

暑假，小金果跟着伯伯到江西考察。

汽车行驶在蜿蜒的山间道路上，一边是绿浪翻腾的翠竹山林，一边是金波滚滚的稻田，景色十分怡人。

停车休息时，看着这迷人的景致，伯伯对小金果说："江西是竹材生产重地，也是稻米之乡。你知道吗，竹子和水稻还是亲缘关系比较近的亲戚呢。"

"竹子像树一样高大挺拔，'稻草'显得弱不禁风。它们怎么会是亲戚呢？难道竹子也是草本植物吗？"小金果疑惑地问。

伯伯回答说，"草本植物大多为一年生，茎秆比较矮小柔软。竹子是多年生植物，茎秆结实高大，木质化程度高，所以竹子不是草。"

"那竹子就是树木咯？"小金果看着伯伯，猜测着说。

伯伯耐心解释道："竹子也不是树木。因为树木不仅有木质部、韧皮部，还有形成层，可以每年长粗；而竹子没有形成层，不能长粗。"

小金果歪着头，有点似懂非懂的样子。

听听竹子的故事

伯伯继续说道:"竹子为多年生常绿植物,有乔木、灌木、藤本,还有呈草本状的,确实不容易归类。"

"那这么说,竹子既不是草也不是树。"小金果试探性地给出了自己的结论。

伯伯赞许地说:"对,非草非木的竹子就是一类非常特别的植物。"

"那请伯伯带着我更好地来认识并了解竹子这一特殊的植物。"小金果热切而认真地说。

森林的故事·竹子篇
The Story of the Forest · Bamboo Chapter

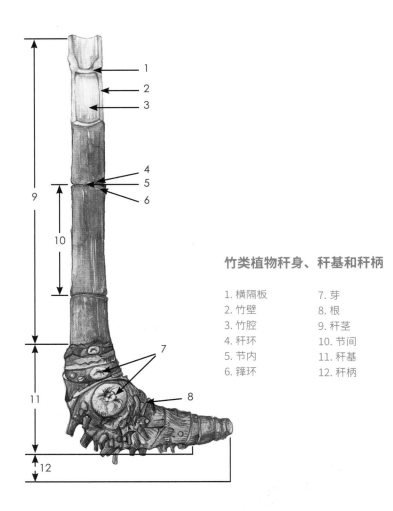

竹类植物秆身、秆基和秆柄

1. 横隔板
2. 竹壁
3. 竹腔
4. 秆环
5. 节内
6. 箨环
7. 芽
8. 根
9. 秆茎
10. 节间
11. 秆基
12. 秆柄

 知识加油站

　　木质素与木质化：木质素是构成植物木质部细胞壁的成分之一。它使木质部维持极高的硬度，以承载整株植物的重量。竹子中有近一半的细胞含有大量木质素细胞，木质化程度高，竹秆坚硬而富有刚性。

竹子开花也结实

春暖花开，草长莺飞。疫情过后的这个春天，小金果特别喜欢和伯伯到公园散步，寻访各种植物的美丽踪影。

亮黄的迎春花、粉红的早樱、雪白的梨花、嫣红的桃花、姹紫的郁金香、艳红的牡丹……这些绚丽多彩的花朵都被小金果记录在他的《植物观察日记》中。通过认真观察，他甚至还记录了松树、枸骨以及木瓜等以前未曾关注过的植物的花朵。但是，有一点小金果觉得很奇怪，虽然仔细寻找，但却从未曾看见一朵竹子花。

今天经过竹林，一系列的疑问再次涌上心头。小金果忍不住问道："伯伯，竹子会开花吗？"

"会的。竹子是种子植物，它也会开花，然后结果。"

"那为什么我从没见过竹子的花呢？"

"那是因为大多竹子的开花周期都非常长，需要几十年甚至上百年才开花，比如桂竹开花就需要120年呢！因而竹子花不常见。"

"那竹子的花长什么样呢？"

听听竹子的故事

森林的故事·竹子篇
The Story of the Forest · Bamboo Chapter

"竹子的花和水稻相似,因为它们都是禾本科植物。它们的花叫颖花。咱们常见的毛竹,花是黄白色的。每朵小花下面有两枚苞片,分别称为外稃和内稃。花瓣已经特化为浆片,因此竹子的花没有艳丽的色彩。一朵至多朵小花加上内、外颖片和小穗轴共同组成低垂的小穗,并着生在竹子的侧枝上。"伯伯一边讲述,一边打开手机图片展示给小金果看。

"细长的花丝悬垂着饱满的花药,很像一串风铃。应该是为了方便它们借助风力传播花粉吧?那它的种子又是什么样的呢?"在给竹花做了个形象的比喻后,小金果接着发问。

"和水稻、小麦一样,毛竹结的果实也属于颖果。由于颖果含有大量的淀粉,因此竹子的果实被称为'竹米',也可以当粮食呢。"

"竹子开花那么少见,那竹米一定十分珍稀。有机会我也要尝一尝。"馋虫上身的小金果,脑海里已经呈现一碗香喷喷竹米饭的情景。

宜兴苦竹的竹花(左)与竹米(右)

知识加油站

种子植物:植物分为孢子植物和种子植物。种子植物都具有开花结实的共性。

森林的故事·竹子篇
The Story of the Forest · Bamboo Chapter

永葆身材的竹子

小金果跟随伯伯去过很多竹乡，比如浙江安吉、江西宜丰、贵州赤水、广东广宁、福建顺昌等。在他的印象中，数不清的竹子几乎一模一样，每一株都通直修长，挺拔多节，青翠欲滴的竹叶惹人喜爱。

这一次来到云南，小金果被路旁一排高大的竹子深深吸引住了，忍不住大声喊道："伯伯，快看，这竹子真粗啊！"

伯伯笑着道："是啊！比小金果的腰还粗。"

小金果双臂环抱着竹秆说："那它一定长了很多年。它是竹子中的'古木'，是'古竹'？"

"这是巨龙竹，它并不古，这株的年龄还不到两岁呢！"

"两年长这么粗？"小金果惊讶了！他突然想起家中院子里的竹子，好像多年都是一个模样，从春笋出土开始就一直保持着苗条的身材。

伯伯看出小金果的疑虑，神秘地笑着说："竹笋有多大，竹子就长多粗。走，我们一起去探寻竹子的生长奥秘。"

伯伯拉着小金果蹲在一根砍伐残留的竹桩旁。"你仔细看，

听听竹子的故事

这竹壁分外、中、内3层。最外侧绿色部分是竹皮,也叫竹青;最里面淡黄色部分称为竹黄;中间部分就是竹肉,它由维管束和基本组织构成。"伯伯轻轻地从竹内壁撕下一片半透明的膜,递给小金果,接着说:"这个可以做笛膜。"

小金果想尽快得到答案,插话道:"为什么每种竹子,不论年龄大小,几乎都一般粗细?"

伯伯瞧出小金果的心思,仍旧慢条斯理地说:"关键就在这形成层。树木能不断长粗,是因为树木的韧皮部、木质部之间具

有分生功能的形成层细胞,它们向外分化不断形成韧皮部,向内分化不断形成木质部。随着分化,树干变大变粗。而竹子的竹秆壁中没有形成层。竹子在笋期,主要靠体细胞的长大来增粗,出土后,它的体细胞生长减慢,所以竹子就很难再增粗了。"

"原来是这样啊!"小金果感叹道,"一点点差别就截然不同,大自然真神奇!"

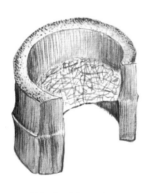

毛竹竹壁的维管束分布

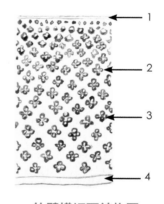

竹壁横切面结构图

1. 竹皮　　2. 基本组织
3. 维管束　4. 髓外组织

 知识加油站

竹子的维管束:竹子属禾本科植物,维管束排列在竹秆髓腔以外和表皮以内,一束一束地分散在薄壁细胞组织内,主要起传输营养和水分的作用。维管束内没有形成层,不能分生新的细胞。

任意行走的竹鞭

伯伯院子的角落里种着几株竹子。一场春雨过后，篱笆墙外传来隔壁邻居惊讶的叫声："我家院子里怎么跑来了一群竹宝宝？"

小金果闻声从屋里跑出去看，只见邻居家院子的草地上冒出了好几个笋头，可爱极了。伯伯也跟过来看个究竟。

"哎呀，都怪我。"伯伯摸着后脑勺，笑着说："对不起！影响到您啦。"

小金果蒙住了，伯伯何错之有？

伯伯找来锹，小心挖开这里的土壤。小金果发现一根细长的"茎"一直延绵到伯伯院子里的那片竹林，它上面还长着好几个笋呢。

伯伯指着说："这叫竹鞭，是竹子的根状茎，节上有芽和不定根，这些芽会长成竹笋或新竹鞭。"

竹鞭是竹子运输和储藏水分与养分的器官，也是繁殖器官。只要季节、雨水、气温合适，它就会在地下随机向各个方向恣意行走，冷不丁还会冒出一根大竹笋。

听听竹子的故事

森林的故事·竹子篇
The Story of the Forest · Bamboo Chapter

小金果急切地问:"那您为什么说怪您呢?"

"你想啊,我们在自家院子种竹子,但没注意控制,导致竹鞭乱跑,给邻居送去了一个意外。所以种植竹子的时候,人们通常要在规定区域周边用水泥板修筑围挡,以防止竹鞭随意生长。"

小金果咯咯笑道:"我知道啦!竹鞭真调皮呢!"

> 有繁殖能力的竹鞭只要条件适宜就会随意生长。

> 竹子为什么会乱跑?

任意行走的竹鞭
Vital Bamboo Rhizome

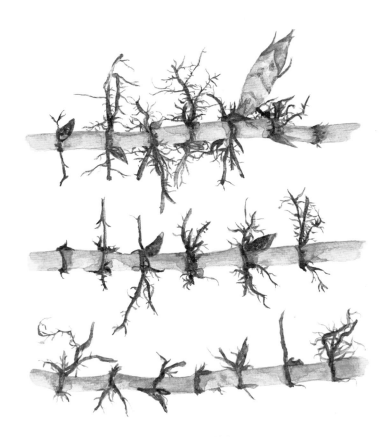

竹鞭形态特征

 知识加油站

  竹鞭：竹鞭是竹子地下茎的俗称，有明显分节，节上生根，节侧有芽，可以萌发成为新的地下茎或发育成笋、出土成竹。竹鞭的穿透力极强，可刺穿坚硬的土壤，甚至可以穿过岩石、砖墙、水泥的缝隙或越过阻碍物形成跳鞭继续生长。

森林的故事·竹子篇
The Story of the Forest · Bamboo Chapter

雨后春笋节节高

渐渐沥沥的春雨接连下了3天,天终于放晴了。

闷坏了的小金果迫不及待地去屋外玩耍。他跑进竹林,走在羊肠小道上,发现地面上冒出了许多竹笋。有的才探出小脑袋,有的却已经蹿得比他还高,高高低低,都裹着深褐色的外衣。

小金果忍不住摸摸这个,比比那个,满脸都写着惊讶。伯伯微笑问他:"是不是好奇,竹林里怎么突然多了这么多的竹子兄弟?"小金果连忙点点头。

伯伯接着说:"有研究人员测量过,出土拔节的春笋,一夜之间可生长近1米。"

"那就是每分钟长高将近1毫米!"小金果认真地算了算,接着问:"雨后竹笋为什么能长得这么快呢?"

伯伯打趣道:"在雨中喝饱了水,在地里憋足了劲,小春笋一出土就迎风生长啦!"

小金果并不满意伯伯的解释,嘟着嘴直勾勾地看着伯伯。

伯伯挑了一个竹笋,挖出来,再从上至下劈成两半,示意小金果仔细观察。小金果发现里面的竹节叠得很紧,好像一个被压

听听竹子的故事

缩的弹簧。

伯伯清清嗓子说:"竹子的生长速度比树木快,是因为树木仅有一个顶端生长点,而多节的竹子在每一节都有一个分生组织。竹笋钻出肥沃的土壤、遇到温暖湿润的天气,它的每一节分生组织都会不断形成新细胞,相邻竹节距就会快速拉长,生长速度也就成倍增长。"

看着小金果迷惑的表情,伯伯进一步解释说:"我们把植物的生长比喻成修建高楼。普通树木的生长就像用传统技术建造一样,只能从下往上、一层一层施工,直至封顶。而竹子的生长就

好像采用现代的框架建造模式进行建造，竹节就是建筑框架，每一个竹节都是一个施工点，所有楼层同时施工，很短的时间里，竹子的大厦就能建好了。"

伯伯一边说一边抬头望向直插云霄的竹梢顶端。小金果也随之远望，并喃喃地说："嗯！我也要像雨后春笋一样快快长大，长得比伯伯还要高！"

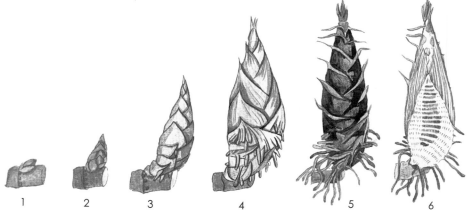

毛竹竹笋的生长

1. 未萌动的鞭上侧芽　2. 萌动的笋芽　3. 初形成的竹笋
4. 出土前的冬笋　5. 刚出土的春笋　6. 刚出土的春笋纵剖面

 知识加油站

竹节与分生组织：竹笋在土中生长阶段，经过顶端分生组织不断进行细胞分裂和分化，形成节、节间、节隔、笋箨、侧芽和居间分生组织，到出土前全笋的节数已定，出土后不再增加新节。竹笋生长从基部开始，先是笋箨生长，继而是居间分生组织逐节分裂生长，推动竹笋向上移动，穿过土层，长出地面。

神奇的竹笋吐水

这是一个云淡星稀的寂静月夜，小金果第一次跟伯伯去竹林开展科研调查。

忽然，小金果仿佛听到了滴滴答答的雨滴声。他抬头望望明月，不解地问伯伯："咦，没下雨呀？可我怎么听到了雨滴声？"

伯伯笑着说："这不是下雨，这是竹笋的吐水声。"伯伯带着小金果蹲下身，在手电筒照射下仔细观察了一番。小金果发现，几乎在每个竹笋上面都挂着亮晶晶的水珠，在手电的强光照耀下犹如一颗颗闪亮的珍珠，有的还正在往下滚落，笋的根部湿漉漉的。

小金果好奇地问："这是竹笋的露珠吗？"

伯伯解释道："它不是露珠，是竹笋吐出来的水。这是竹子特有的吐水现象。竹笋快速生长需要大量养分，而养分通过水分进行运输。于是啊，竹子的根系就像装了个小水泵，将土壤中的水分和养分大量吸收上来，源源不断地输送给竹笋。养分被竹笋快速吸收，而多余的水分就会顺着箨叶吐出来，湿润幼竹周围的表层土壤，这样就形成了竹林内的水分小循环！"

听听竹子的故事

森林的故事·竹子篇
The Story of the Forest · Bamboo Chapter

伯伯顿了顿,接着说:"尤其是在温度高、湿度大的夜晚,竹子渗出的水分更多,形成的水珠也更大。无数水珠滴在地面干枯的竹叶上,就好似滴滴答答的雨声。"

小金果听后,恍然大悟道:"夜里竹林里滴滴答答的雨声,原来是竹笋在玩吐水呀。哈哈!"

神奇的竹笋吐水
Magical Phenomenon - Bamboo Spitting Out Water

竹笋吐水

知识加油站

竹笋吐水：通常发生在气温高、湿度大、空气中水蒸气接近饱和的夜晚。此时，竹笋的蒸发量减少，但根系仍然猛烈地吸水，导致体内水分吸入量大于蒸发消耗量，于是过多的水分就会从箨叶尖排出，形成水珠。这种吐水现象遇上高湿高温的夏季，也反映在竹叶叶尖上。简而言之，就是根压造成吐水。

森林的故事·竹子篇
The Story of the Forest · Bamboo Chapter

竹笋竹子是同辈

春风吹过春雨落,竹林里长出了许多的竹笋。小金果和伯伯去竹林挖竹笋。小金果兴奋地摸摸这个,碰碰那个。笋芽儿尖尖的,硬硬的,像一个个愣头愣脑的小娃娃。

小金果问道:"伯伯,竹笋是竹子的孩子吧?那笋应该是竹子的种子喽?"

"不是的。"伯伯一边回答,一边用铁锹拨动铲除地面上的枯叶和浮土,不一会儿就露出了一片地下网络——蔓延纵横、连成一体的地下竹鞭。

"竹笋是竹子地下茎上的芽,也就是竹鞭的芽。竹笋生长成为竹子。"指着竹鞭上的一个竹笋,伯伯接着说,"竹子会开花结果、繁育下一代,但大多数竹子开花所产生的种子是发育不完全的,仅有少数竹种能够结出有生命力的种子。"

"噢,我明白了,竹子开花结的种子才算是竹子的孩子。"小金果若有所思地说。

"有趣的事情来了!"伯伯顿了顿说,"这些新长出的和即将长出的笋或竹,严格地说不算是这片竹林的'下一代'。如果

听听竹子的故事

我们把竹林比喻成一棵树,那它们更像是新长出来的新芽和新枝。"

小金果似乎一下子明白了,说:"哦——无论地面上的竹子是老是幼,只要是在同一竹鞭,它们就是同根生的同辈竹。"

竹子与竹笋的形态特征图

 知识加油站

  竹林与竹树：根据植物学观点，地下相连的一片竹林或竹丛，地下茎是"竹树"的主茎，竹秆是"竹树"的分枝，竹笋就是"竹树"的枝芽，它们共同组成了"一棵树"。

赏竹
青青四季同

森林的故事·竹子篇
The Story of the Forest · Bamboo Chapter

竹中之最

伯伯领着小金果在绿博园游览。绿博园里的竹子种类真丰富，有飘逸潇洒的凤尾竹、形态奇异的龟甲竹、憨态可掬的佛肚竹、通体金黄的黄秆竹、珍贵奇特的花毛竹、优雅美丽的斑竹……多姿多彩。

小金果看见前方有一片草坪，欣喜地嚷着要上去打个滚儿。跑近一看，发现这些草与众不同。它们匍匐在地，线状披针形叶片分两行排列，节处密被细毛。

看见小金果惊异的眼神，伯伯顺手牵起这样一根"草"说："这不是草，而是世界上较矮小的一种竹子，高不盈尺，名叫无毛翠竹。"

听听竹子的故事

小金果来了兴趣，脱口问道："这是最小的竹子，那最大的竹子呢？"

伯伯一脸难不倒的样子，说："世界上最大的竹子是巨龙竹。它高大魁梧，最高可达十几层楼高，有成人大腿那么粗，被称为'竹王'，是我国珍稀特有竹种，分布在云南省西南部。"

小金果摇着伯伯的胳膊说："伯伯带我看过巨龙竹。"

伯伯连连说："我们还要看世界上最细的竹子、最粗的竹子；竹节最长的竹子、最短的竹子；竹壁最薄的竹子和实心竹子……"

小金果心驰神往，开心地笑了。

> 竹子种类繁多，我国就有600多种。

> 竹子真是千姿百态啊！

森林的故事·竹子篇
The Story of the Forest · Bamboo Chapter

无毛翠竹形态特征图

 知识加油站

　　竹类植物： 竹类植物物种丰富，世界上有竹类植物90余属1400余种。有高几十米、直径几十厘米的乔木状竹子，也有高度只有几十厘米、直径几毫米的草本状竹子。我国竹子种类繁多，全国有40余属600余种。

竹门三家族

中国竹子博物馆是国内唯一的一家竹子专业博物馆。在这里不仅可以一览世界各地的奇篁异筠，还能了解竹子的千载加工利用史。

小金果置身其中，目不暇接。看着千姿百态的竹子，他脑海中闪现出伯伯曾经提到的"竹门三家族"这个词，对此他一直不解，于是着急地向伯伯提问。

伯伯没有立刻回答，而是牵着小金果走到竹根标本展柜前让他观察。

小金果说："我知道，这是竹鞭，也就是竹子的地下茎。"

"对的。你再仔细看，这3个样本就是竹子地下茎的3种形态：单轴型、合轴型与复轴型。人们通常以植物的生长特点来鉴别种类，对应竹子地下茎的这3种形态，将竹子分为三大家族——散生竹、丛生竹和混生竹。"

"那，这种竹子属于哪一种呢？"小金果指着一个标本问道。

"这是单轴型的地下茎——形成散生竹。它的竹鞭细长，在土壤中横向蔓延生长。竹鞭的节上长芽，鞭芽发育成笋，出土成竹，

听听竹子的故事

森林的故事·竹子篇
The Story of the Forest · Bamboo Chapter

在地上形成稀疏散生的成片竹林。"

"那这个是不是合轴型?"小金果指着一个标本说,"我见到一丛一丛的竹子,它们抱团抱得可紧了,根本走不进去!"

"对!"伯伯高兴地说:"合轴型地下茎形成丛生竹。它的竹鞭极度缩短,竹节很密,因顶芽出笋长成的新竹都紧挨着老秆,

从而在地上形成竹秆密集生长的现象。"

"那复轴型又是什么样的呢?"小金果紧接着问。

伯伯引金果走到另一个标本前,解释说:"复轴型地下茎兼具单轴型和合轴型地下茎的特点,所以地上部分既有竹秆密集的竹丛,也有稀疏散生的竹秆,两者共同形成成片竹林。这类竹子就称为混生竹。"

"嗯,这下我总算搞明白'竹门三家族'是怎么回事了!各式各样的竹子原来可以归属于3个家族,这样好记多啦!"

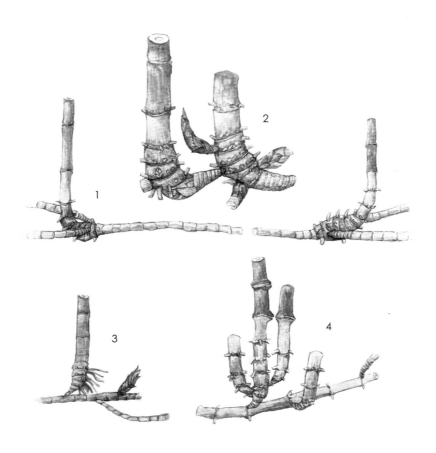

竹类植物的地下茎类型

1. 合轴散生亚型 2. 合轴丛生亚型 3. 单轴型 4. 复轴型

 知识加油站

　　竹子地下茎：竹子地下茎可分为单轴型、合轴型与复轴型三大类。单轴型地下茎形成散生竹；合轴型地下茎形成丛生竹；复轴型地下茎形成混生竹。

延绵似藤的攀缘竹

小金果与伯伯在西双版纳热带植物园参观。小金果看见路边树林中的高大乔木上倒挂着无数根垂帘,一节一节的,长着青翠的细叶。小金果兴奋地说:"把这些藤蔓编织起来,我们就可以做个天然的秋千了。"

伯伯笑着说:"这可不是平常热带雨林中普通的藤本植物。它们是竹子。"

"怎么会是竹子呢?"小金果疑惑极了,在他印象中,竹子可都是通直挺拔的模样。

伯伯领着小金果走进林内仔细观察。只见这些竹子三五一丛,直径为1~3厘米。秆长有二三十米,每节分支数十条短枝,簇生于节部。它们有的攀附着乔木直至树梢,而后悬垂于林冠外;有的匍匐在地,延绵成片。

看着小金果不解的神情,伯伯解释说:"这是竹类中外形独特的一个种类,具有观赏价值。它们叫'藤本竹'。"

"由于外形像藤,所以命名为藤本竹?"

"是的。你能找到它与你平时看到的藤本植物在外形上的最

听听竹子的故事

大区别吗?"

爱挑战的小金果没有退缩,他认真地打量着眼前这一丛竹子,试探着回答:"它们秆体光滑,没有小触手。"

"小金果不愧是个小科学迷,观察够仔细!"伯伯竖着大拇指说:"就像你说的,藤本竹不具有卷须一类的攀缘器官,所以

它是依附树干向上生长的。当到达树冠枝端缺乏倚靠后,它的梢部就会悬垂下来,直至地面。所以有些地区的人们也称它为'吊竹'。"

"中国藤本竹类有藤竹属、梨藤竹属、长穗竹属、悬竹属和新小竹属5属20种。另外,还有空竹属、薄竹属、箭笢竹属、泡竹属和单枝竹属等约15种半攀缘状竹类。它们大都分布在热带及亚热带高温多雨的低海拔河谷地区,云南是中国藤本竹类分布最集中的地区。此外,在海南和西藏也有,它们被归集为琼滇攀缘竹区,是中国五大竹区之一。"

"五大竹区之一。那另外的四大竹区是哪些?"

"这个问题就留给你自己去寻找答案了!"伯伯慈爱地拍着小金果的脑袋,给他布置了一道课后作业。

知识加油站

竹子的分布:中国是竹类中心产区之一,竹子分布具有明显的地带性和区域性。按竹类植物的分布情况,划分为北方散生竹区、江南混生竹区、西南高山竹区、南方丛生竹区和琼滇攀缘竹区5个分布区。

西藏新小竹形态特征图

1-2. 秆的一部分，示分枝　　3-4. 秆箨的腹面及背面

腰肥肚圆的佛肚竹

伯伯近日搬回一盆竹盆景，十分欢喜，每天呵护之际都忍不住吟唱："宁可食无肉，不可居无竹。无肉使人瘦，无竹使人俗。"伯伯笑呵呵的模样，像极了弥勒佛，也像极了眼前的竹子。

小金果凑近观察，这盆栽中的竹子形态奇特，竹节较细，可节间却短粗膨大，好似弥勒佛的肚子，又好像叠起的罗汉。小金果被这滑稽可爱的样子逗笑了，忙拉着伯伯问："这是什么竹子？大肚子竹吗？"

伯伯更加得意，冒出了京腔："这就对啦，它就是珍贵的佛肚竹啊。"

"我想知道，它为什么会长成这样啊？"小金果一本正经地提问，打断了伯伯的自我陶醉。

"通常竹秆生长时，同一节上细胞分裂和伸长活动基本上是齐头并进的，所以节间形状大多为正常的长筒形。但佛肚竹同一节上的细胞分裂与伸长却不同时、不同速，这就导致竹秆节间的外形畸形变化，或者竹节交互倾斜形成奇特的形态，也由此成为竹间'怪物'，用于园林造景赏玩。"

听听竹子的故事

森林的故事・竹子篇
The Story of the Forest · Bamboo Chapter

小金果挠挠头，得出结论："这是不是'无心插柳柳成荫'呢？"伯伯为小金果的机智总结拍手称赞。

大肚子竹子是一个种吗？

畸形生长形成奇特形态。

腰肥肚圆的佛肚竹
Bambusa ventricosa with Fat Waist and Round Belly

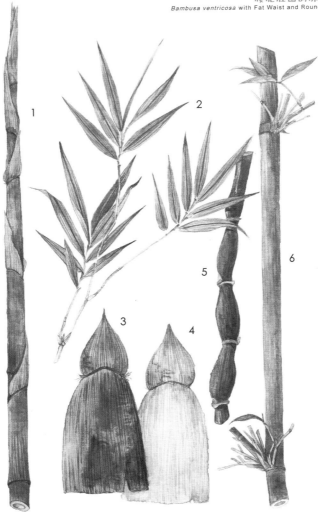

佛肚竹

1. 笋 2. 枝叶 3. 秆箨的外表面 4. 秆箨的内表面 5. 异形秆 6. 正常秆

 知识加油站

　　竹节的畸形生长：竹子生长过程中，当细胞分裂、伸长和加大速度不同时、不同速时，秆壁内的细胞形状和排列就会产生畸形，从而引起竹秆外形变化，形成畸形的节间溢缩或肿胀，或竹节发生交互倾斜。

森林的故事·竹子篇
The Story of the Forest · Bamboo Chapter

嵌金镶玉的竹子

南京林业大学的白马基地竹种园，是竹类科研教学、学术交流示范和旅游观光休闲的基地。

竹种园内曲径通幽。伯伯边走边滔滔不绝地介绍园内丰富多样的竹品种，小金果则边听边四处观察。

突然，他被左前方的一片竹子吸引住了目光。这竹子秆形通直，色彩夺目，犹如根根金条上镶嵌着块块碧玉。

没等小金果发问，伯伯就开口介绍道："这是金镶玉竹。竹如其名，它的竹秆为黄色，在每节生枝叶处都天然生成一道碧绿色的浅沟，位置节节交错。你可以上前去验证一下。"

小金果满腹狐疑，钻进竹林，一根一根观察。望着小金果稚气的样子，伯伯说："新发的金镶玉竹竹秆为嫩黄色，而后渐变为金黄色，节间形成绿色纵纹，黄绿相间。"

"它们为什么会长成这样？"小金果发问。

"这和竹叶与竹秆中的叶绿素、类胡萝卜素和花青素等物质的含量有关，它们就像是不同的颜料，这些不同色彩颜料的调和造成了叶片与竹秆色彩的变化。类似的还有那边的黄槽竹，你看，

听听竹子的故事

它们是绿色的竹秆上镶着金色的条纹。再远处的紫竹，它们的幼秆是绿色，一年后竹秆渐渐出现紫斑，最后全身变为紫黑色。"

小金果赞道："小小竹秆竟有这般魔幻变化，看来竹子也是一位高明的调色师啊！"

森林的故事·竹子篇
The Story of the Forest · Bamboo Chapter

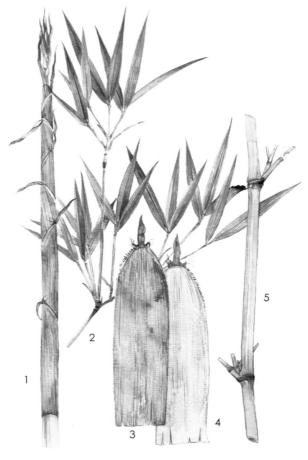

金镶玉竹

1. 笋　2. 枝叶　3. 秆箨的外表面　4. 秆箨的内表面　5. 秆

 知识加油站

竹秆中的色素变化：植物叶、秆、果所呈现的红、绿、黄色彩，往往与它们所含的叶绿素、类胡萝卜素和花青素色素的类型与含量有关。研究表明，与全绿竹秆比较，花秆竹子控制色素的基因发生改变，影响了竹秆叶绿素的沉积，从而产生花秆。

母慈子孝的孝顺竹

小金果和伯伯坐游轮去四川。清晨，站在船头倚栏眺望，但见薄雾弥漫着岷江两岸的田野，一丛丛的绿植，或近或远，亦浓亦淡，美得超凡脱俗。船靠岸后，他们来到一个度假山庄用餐。

走进庭院，只见叠石假山边种植着一丛竹子，叶子密集下垂，婆娑秀丽，与船上远观植物的姿态一样，小金果于是连忙请教伯伯。

伯伯富含深情地说："这是孝顺竹，也叫凤凰竹、慈孝竹。"

"它的竹丛形态像美丽的凤尾，因此取名凤凰竹。那名叫孝顺竹，是不是也有它的特殊含义呢？"小金果紧接着问道。

"孝顺竹是丛生竹。它最显著的一个特征是竹鞭紧密团结，新笋围绕母竹生长，形成一丛一丛的生长方式，竹鞭不随意扩张，竹笋不无序生长，而且这种生长方式永不改变。"

伯伯看小金果好像没听明白，又接着解释道："这种竹子发出的新笋就像孝顺的孩子一样地紧紧围绕在母竹身边，而母竹又像慈祥的母亲一样永不嫌弃笋子的累赘，母慈子孝，共同成长。因此人们给它取了个温馨的名字——孝顺竹。"

听听竹子的故事

森林的故事·竹子篇
The Story of the Forest · Bamboo Chapter

小金果忽然想起合轴型的知识点，举一反三："所以孝顺竹应该是合轴型吧！这名字真配它！"

"对，对！"伯伯连连赞许，"有关孝顺竹，民间还有很多有趣的故事呢，你有空可以找来看看。"

"我也要像新笋一样，永远爱护妈妈、保护妈妈。"小金果在心里默默地说。

母慈子孝的孝顺竹
Bambusa multiplex

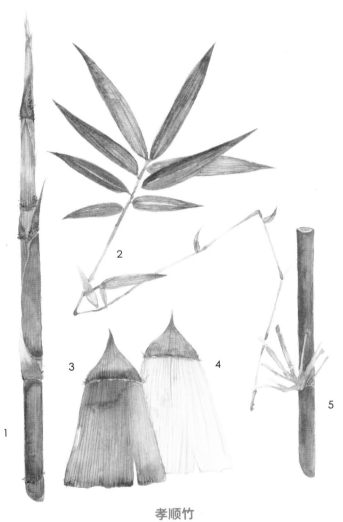

孝顺竹

1. 笋　2. 枝叶　3. 秆箨的外表面　4. 秆箨的内表面　5. 秆与分枝

 知识加油站

丛生竹：丛生竹的地下茎不是横走地下的细长竹鞭，而是极度缩短，节密，因顶芽出笋长成的新竹都紧挨着老秆，从而在地上形成竹秆密集生长的竹丛。

森林的故事·竹子篇
The Story of the Forest · Bamboo Chapter

身洒千滴泪的斑竹

伯伯珍藏了一支竹笛，时常拿出来把玩。被摩挲得锃亮的竹笛表面点缀着不规则的斑点，显得很别致。

小金果禁不住问："伯伯，这支笛子是不是非常贵重？它是用什么竹子做的呢？"

伯伯说："笛子虽然是普通的乐器，但这支笛子的确比较珍贵，因为制作它的竹材很珍贵。它是用斑竹制成的，斑竹也称为湘妃竹。"

小金果盘算着伯伯的话：斑竹，看外形就能明白；湘妃竹？恐怕又有故事了吧。

伯伯介绍说："斑竹分布在黄河和长江流域各地。据说岳阳君山岛的斑竹上有云纹紫色斑迹，很像泪痕。若将斑竹移栽别处，第二年斑迹就消失得无影无踪，如果再将这株斑竹移回君山，下一年又是斑痕累累的了。"

"啊，这么神奇！那这斑痕从哪儿来的呢？"小金果好奇地问。

"竹类专家研究发现，斑竹身上的斑原来与它生长的土壤、

听听竹子的故事

身洒千滴泪的斑竹
Phyllostachys bambusoides 'Lacrimadeae'
Sprinkled with Thousands of Tears

气候条件密切相关，它们的花纹实质上是细菌侵蚀竹身后在竹子表面形成的各种菌斑花纹。新竹秆起初并没有斑点，当长到九个月时，才渐渐长出斑痕来。"

屏住呼吸的小金果轻舒了一口气，说："哦，原来如此，那它为什么又叫湘妃竹呢？"

"相传尧舜时代……"伯伯故意拖长声音卖起了关子，接着说："你可以自己上网查看一下这个神话故事哟。"

小金果假装嘟起了嘴，却快速地打开电脑，去查资料啦！伯伯欣慰地笑了。

森林的故事·竹子篇
The Story of the Forest · Bamboo Chapter

斑竹

1. 竿　2. 枝叶　3. 秆箨的外表面　4. 秆箨的内表面　5. 笋

 知识加油站

　　湘妃竹：湘妃竹又名斑竹。竹秆布满褐色的云纹紫斑，是著名的观赏竹。秆可用来制作工艺品，亦可制作竹材。

种竹

无竹令人俗

森林的故事·竹子篇
The Story of the Forest · Bamboo Chapter

竹子种植讲科学

植树节到了。小金果和竹类研究所的叔叔阿姨们去生态园种竹子。

坐在车上,小金果浏览着沿路春光,一边思考一边问:"伯伯,竹子要怎么种?是不是和种树一样种下小竹苗啊?"

伯伯咂着嘴饶有兴趣地说:"种竹和种树有相似点,也有不同。"看着小金果专注的神情,伯伯接着说:"种竹子有几种方法。第一种是种子育苗,就是把竹种子撒在苗圃里,长成竹子后进行移栽;第二种是埋鞭育苗。就是在早春或秋季,将挖取的健壮竹鞭埋在平整后的土地里。"

"那我们今天采用哪种方法啊?"小金果急切地打断了伯伯的话。

伯伯正准备回答,车已到了目的地。下车后,小金果看见空地上摆放了许多根部用草绳捆扎成球形的竹子,忙拉着伯伯上前看个究竟。这些竹子青翠欲滴,节间匀称,有两个手指那么粗,但都被截去了顶梢。

小金果心里犯嘀咕,扭头问道:"伯伯,我们要种的竹苗和

听听竹子的故事

竹鞭呢？"

"今天我们采取第三种种植法。"伯伯笑着说："这种方法就是选取生长健壮、无病虫害、大小适中的竹子作为母株来种植。"

说干就干，小金果和伯伯忙碌起来。伯伯边干边补充道："散生竹适用分株繁殖的方法，丛生竹适用分蔸繁殖的方法。"

小金果起劲地干啊干，他累极了，不由地直起腰，抬手拭去额角的汗珠。看着眼前一排排刚种的竹子，仿佛望见了来年春天出笋、夏季成林的景象。他开心地笑了。

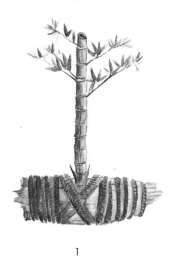 

母竹的移栽方式

1. 母竹的运输包扎　2. 母竹栽植后架设支架

 知识加油站

　　母竹移栽：母竹移栽是竹子种植的主要方法之一，分4个步骤：第一是选好母竹，以壮龄期无病虫害的竹为佳。通常，毛竹选2～3年生竹，其他选1～2年生竹。二是挖好母竹，需连竹秆带竹鞭挖出。三是运输母竹。运输前应截去竹梢，只留最下层3～5盘竹枝。这既便于运输，也可减少母竹水分消耗，有利于保护母竹。四是科学栽竹。栽植时保持鞭根舒展，覆土深度比母竹原入土部分稍深3～5厘米。应设立防风支架，以防竹秆摇晃。

科学克隆小竹苗

这里是竹子研究中心。伯伯领着小金果去参观竹子组培实验室。

换上大白褂,走进洁净的实验室。只见一排排金属架子上整齐地排列着一个个小玻璃瓶,每个瓶子里都长着翠绿的小竹苗。小金果看着这一瓶瓶小绿苗,觉得非常新奇。

伯伯介绍说:"这些小竹苗都是使用克隆技术培养出来的。"

"什么,竹子也能克隆?"小金果问道。

伯伯说:"克隆技术已经运用在竹子培育中。人们挑选优良的竹子,采用它们的芽、花或种子,经无菌消毒后,接种在配制好的培养基里,然后放到人工控制好温度、湿度及光照等条件的环境中进行培养,最终长成有芽有根的完整小竹苗。"

听听竹子的故事

"这么神奇啊!大多数竹子开花结实周期长,克隆竹苗比竹种育苗要省时快捷多了!"

伯伯说:"是啊,用这个方法生产竹苗,就不会受季节的限制了,而且具有产量高、不占用耕地等优点。科学家利用新技术,通过改变竹子基因,还能培育出我们需要的新品种。"

小金果拍手称奇:"科技促进生产。竹子家族又有新成员喽!"

竹子组织培养的形态特征

1. 芽萌动　2. 芽增殖　3. 芽生长　4. 生根

知识加油站

植物克隆：植物的枝、叶、芽等部分器官或组织包含了整个植株的所有信息。通过控制技术提供适宜的温、光、气、热、营养、激素等，对植物的枝、叶、芽等部分器官或组织进行培养，实现快速成苗，这就是植物克隆。

森林的故事·竹子篇
The Story of the Forest · Bamboo Chapter

竹笋采挖有技巧

正是冬笋肥美的季节。听到伯伯说要去挖冬笋，小金果兴奋地跳了起来。连忙和伯伯准备好锄头、铲子、砍刀后，就开心地出发啦。

小金果边走边问："竹笋分为春笋、冬笋，是不是因为生长在不同的季节而取名的？"

听听竹子的故事

"是的。除了冬春笋,还有夏笋。但冬笋味道最佳,素有'金衣白玉、蔬中一绝'的美誉。"

小金果听后不由得寻思起来:挖笋,会不会减少来年的出笋量和新竹量。他急忙向伯伯求证。

伯伯解答道:"放心吧,我们挖笋是有规定的:新种植毛竹3年内禁止挖冬笋;3年后,当每亩毛竹达180株以上时才可以有计划地挖笋。合理的采挖不仅可以增加林农的经济收入、满足消费者的需求,还可以保证竹子的正常生长。"

森林的故事·竹子篇
The Story of the Forest · Bamboo Chapter

小金果如释重负，轻快地哼起了小调。

伯伯在竹子周围仔细观察，看到一处地表有裂隙的地方，用脚轻轻地踩了踩，然后拿起锄头慢慢开挖，露出笋尖后接着用铲子剔除附着的土壤，最后用砍刀小心翼翼地取出冬笋。

见伯伯做得细致认真、有条不紊，小金果问："这么麻烦，为什么不把土壤都挖开找笋呢？"

"我们刚用的是开穴挖笋法。你说的是全面翻土挖笋法，通常结合竹林冬季的松土施肥，对竹林进行抚育垦复时挖掘冬笋，以中翻20厘米左右为宜。有经验的竹农还会采用沿鞭翻土的挖笋法。但不管哪种方法，都要保护竹子的根鞭不被损伤与折断。"

小金果认真地点了点头，细心地配合着。经过一阵劳作，他们收获了一篮冬笋，满载而归。

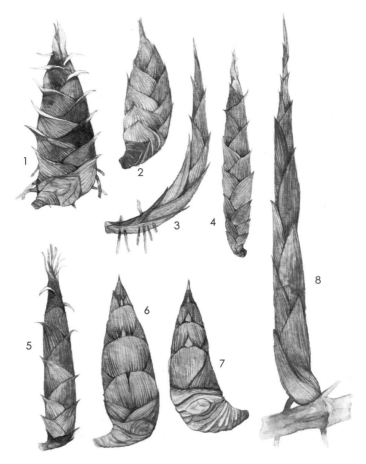

几种竹笋的形态特征

1. 毛竹春笋　　2. 毛竹冬笋　　3. 毛竹鞭笋　　4. 刚竹笋
5. 淡竹笋　　　6. 麻竹笋　　　7. 绿竹笋　　　8. 方竹笋

 知识加油站

　　冬笋挖取与竹林抚育：冬笋通常指尚未出土的毛竹笋。毛竹竹鞭上每个节上都有芽，每个芽的生理状态都不同，在自然状态下能够发育成竹笋的不超过10%。挖取冬笋不仅可以获得美味山珍，也有利于促进其他笋芽的分化，不会影响竹林的立竹度和整齐度。尤其是在冬季结合挖冬笋实施竹林垦复，既挖了冬笋，又抚育了竹林，促进竹林经济效益和生态效益的提高。

 森林的故事·竹子篇
The Story of the Forest · Bamboo Chapter

毛竹标号建档案

听听竹子的故事

盛夏，走进郁郁葱葱的毛竹林，小金果顿时感觉到一丝凉爽。他抬头仰望，繁茂的枝叶犹如一把碧绿的巨伞，遮挡住了毒辣辣的太阳，给林间投下一片清凉；他举目远望，一株株翠竹绿得流油，

在斑驳的光影中犹如林间跳动的音符。

突然，小金果发现那翡翠般的竹秆上有些红色的印痕。走近一看，是红漆标记的数字"8"。小金果觉得奇怪："有人不爱惜竹子，用刀在上面刻名字。而这红漆数字又是谁写的呢？"

他仔细地环视了一下四周的竹子，发现很多竹秆上都写着不同的数字，除了"8"，还有"9""7"等。他纳闷地问："伯伯，竹子上的数字代表什么？神秘符号吗？又是谁在竹子上写数字的呢？"

伯伯乐了："这是工人叔叔们做的标记。对于这种大型林场，为了方便严格控制毛竹砍伐的年龄，每到七八月份，林场工人都会在竹秆上标上它出笋的年份。如果是第一年实施，通常要对竹林的每一株竹子都进行年份编号，称为全竹号字。有了这些数字，每一棵竹子就相当有了自己的个人档案。工人在采伐时也就心中有数啦。"

"7，8，9"小金果喃喃着，脑海灵光一现："我知道啦！这是年份对不对！那个8就代表它是2018年发笋长成的。对吗？"

伯伯鼓掌："真聪明！有了这个标号，工人按号砍伐，就能保证竹林合理的立竹年龄结构。再坚持'砍老留小、砍密留稀、砍弱留强'的砍伐原则，竹林就能健康生长、一直保障利用啦！"

小金果笑道："聪明的不是金果，是竹子经营管理专家呀！"

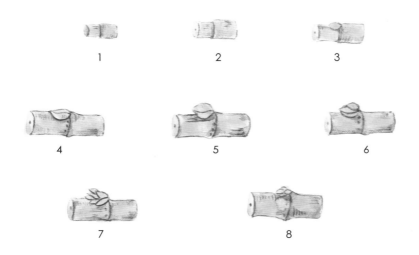

毛竹的竹鞭年龄和侧芽的变化

1-2. 鞭梢部分的鞭节　3-4. 幼龄（1～2年生）竹鞭的鞭节
5-6. 壮龄（3～6年生）竹鞭的鞭节　7-8. 老龄（7年生以上）竹鞭的鞭节

 知识加油站

　　竹子的年龄：竹子的年龄由竹龄、鞭龄和叶龄组成。竹秆的年龄称为竹龄。一株竹子从竹笋出土形成新生竹，经历幼龄、壮龄和老龄，直至老死，一般历经十几年。

竹子钩梢抗雪灾

金秋十月,小金果和伯伯一行到石塘竹海度周末。坐在茶室小憩,小金果发现湖对面山坡的竹林在大幅度地摇摆起伏。定睛一看,几人正在钩弯竹秆,并截断竹梢。

"有人在破坏竹林!"小金果急得大喊起来,手指向前方。

伯伯顺着金果手指的方向望去,侧头安慰金果说:"别着急!

听听竹子的故事

他们不是在破坏,而是在治理,是在给竹子钩梢呢。"

"为什么要这样做?"小金果问。

"因为毛竹枝叶繁茂,如果冬季遭受大雪,枝叶上就会积存大量的冻雪和冰挂,使得毛竹头重脚轻,进而发生翻蔸、断裂、倒伏等灾害现象,这对竹林生长和产量会产生很大的不利影响。钩梢能预防并减少这种隐患。"

小金果还是不放心,让伯伯领着他到跟前去看个究竟。只见护林工人手持锋利的钩梢刀,对准新竹顶端,快速地削掉竹秆顶端的梢头和侧枝。

伯伯接着说:"毛竹钩梢防灾又增产。去梢的竹子,极大地增强了抗风、防冻、抗雪压的能力。钩梢之后,竹林的透光性更好,光合作用更强,有利于促进竹叶、竹鞭和竹笋的生长,使产量提升,而且竹秆也更加挺直。"

听完伯伯的解释,小金果欣慰地笑了,同时也为自己的不知情对人的误解有点儿脸红。

  知识加油站

竹林钩梢:钩梢是针对当年出笋长成的新竹,用快刀钩去竹秆上的枝梢,其主要作用是减轻冬季与早春的雪压之害。钩梢按操作季节分3种:一是霉梢,即在6月黄梅天进行;二是伏梢,在7~8月大伏天进行;三是白露梢,在9月初白露前后进行。

用竹
不可居无竹

森林的故事·竹子篇
The Story of the Forest · Bamboo Chapter

一竿为舟的竹

暑假，小金果和伯伯到贵州旅游。

"哇！伯伯快看！"小金果突然惊奇地大喊。顺着小金果手指的方向看去，只见一名男子脚踩着一根粗大的竹子，持一根竹竿做桨，在水面上缓缓地划行。"噢，这是赤水河流域著名的独

听听竹子的故事

竹漂绝活,当年红军四渡赤水时就曾以独竹做舟渡河。"伯伯笑着说。

"一根竹子怎么就能支撑一个人在水面漂流?"好奇的金果提出了疑问。伯伯笑着摸了摸小金果的头,没有立刻回答。

伯伯领着小金果来到河边,河滩上放着一根大毛竹。"这就是我们刚才见到的独竹漂所用的竹子。"伯伯说。

小金果蹲下仔细端详,估摸着这根竹子的大头有自己的大腿

森林的故事·竹子篇
The Story of the Forest·Bamboo Chapter

那么粗,长度约两层楼高,除此之外好像也没什么特别之处。这就能承载一个人的重量?

伯伯也蹲下身子,提示小金果:"你数数这根竹子有多少节?"小金果欣然行动。

伯伯接着说:"竹子每节都是密封的空室,就像一个个小浮球,产生的浮力足够托起一个人。"话锋一转,伯伯提问道,"你是不是在想,是不是什么竹子都可以独自成舟呢?"

"不可以吧。我想一定要像这根竹子一样,要足够粗、足够长才行。"小金果因仔细观察了一会儿,才试探着回答。

伯伯满意地点点头说:"答对了!竹子如果细了、短了,浮力自然就减少了,也就托不起一个人啦。"

伯伯接着补充说:"正因为这个地方多产这种大竹子,独竹漂也就成为当地的一项民间绝技。在2011年的少数民族传统体育运动会上,独竹漂第一次被列入运动项目。"

"好厉害!"小金果眼里闪耀着钦佩的光。

 知识加油站

竹秆结构:竹秆是竹子的主体,包括秆柄、秆基和秆茎3个部分。秆茎由秆环、箨环、节内、节隔和节间组成。秆环、箨环和节内称为节。两节之间称节间。节间通常中空,节与节之间有竹隔相隔。不同竹种竹秆的大小、节数、节间长短等有显著差异。

一竿为舟的竹
One Pole as a Boat

毛竹的形态特征

1. 竹笋　2. 枝叶　3. 秆箨的外表面　4. 秆箨的内表面　5. 竹秆

森林的故事·竹子篇
The Story of the Forest · Bamboo Chapter

置景园林的竹

听听竹子的故事

跟随伯伯天南地北走过许多地方，欣赏了许多园林，小金果发现到处都有竹子的身影。他问伯伯："园林中人们为什么喜欢种竹子啊？"

人们为什么喜欢在园中种竹子？

也许是问题太大,伯伯思索了一会儿才说道:"中国是'竹子王国',竹子种类繁多,分布广泛,容易利用。"

受伯伯话的启发,小金果抢着说:"翠竹青青,千姿百态,集形态美、色彩美、意境美于一身,观赏价值高。"

伯伯竖起大拇指,也学着小金果的口吻说:"是啊!不同竹种形态各异。有的盈不足尺,有的高耸云天;有的茕茕孑立,有的聚拢成丛;有的叶细似针,有的叶大如掌……"

竹子种类繁多,形态各异,具有造景观赏价值。

森林的故事·竹子篇
The Story of the Forest · Bamboo Chapter

伯伯说话越发文气了:"以竹造园,不论其本身形成的纷披疏落竹影,还是以竹造景、借景、障景,或是用竹点景、框景、移景,都能组成一幅幅如诗如画的美景,风格多式多样。"

"我来说,我来说!"小金果也兴致高昂,打开了话匣子,"庭院可种观秆、看色、赏形的竹子,如佛肚竹、金镶玉竹;分隔空间可以种植高大的毛竹;修饰角落可以选用常绿的凤尾竹、淡竹、紫竹、龟甲竹等;竹篱嘛,用丛生竹、混生竹最合适。"

伯伯激动得连连说:"带你跑了这么多地方,真没白跑!"

知识加油站

中国园林与竹子:中国人十分喜爱竹子四季常青、不畏严寒、节高心虚的性格,提倡把竹子作为做人的标准。园艺工作者根据竹子秆、枝、叶、笋的形态进行选育观赏,创造了中国竹子园林的特色。

钢筋铁骨的竹

这是一处美丽的傣族村寨。一座座独特的傣家竹楼,掩映在一丛丛浓密的凤尾竹与油棕树林中,显得那么宁静恬美。导游小姐介绍:傣族地区是竹材的盛产地。村民们喜欢就地取材,用粗竹子做龙骨搭框架,用竹编篾子做墙体,用竹篾或木板做楼板,再在屋顶上铺草,一座结构简单、通风透气的竹楼就建成了。

离开竹楼,小金果问伯伯:"为什么随风飘摇的竹子,竟然能造出两层小楼,住人、储物、养牲口,一点儿不耽误?"

"这都是因为竹子有着特殊的'钢筋铁骨'身板。"伯伯一边走一边和小金果聊。

"竹子的特殊身板源于它的生长方式。竹子由外向内生长,因此表层最硬;竹子一般长得很高,但茎秆底部只比顶端略粗。茎秆能承受自身重力,并具有高抗拉强度。这样它们才能既随风弯曲,又不会折断。折断竹秆,你会发现有很多不容易扯断的细丝,那就是坚韧的竹纤维,它们把整个竹秆紧紧连接在一起,就好比造高楼用的钢筋。"

小金果问道:"那竹子的空心、多节结构,产生什么影响呢?"

听听竹子的故事

森林的故事·竹子篇
The Story of the Forest · Bamboo Chapter

"竹子空心结构使它的抗弯能力比相同横截面积的实心结构要大很多;竹节又很像坚硬的'横梁',起着支撑作用,层层加固。"伯伯笑呵呵地说:"这样,你明白了吧。竹材是一种强度高、稳定性好的建筑材料呢!"

小金果听了,羡慕之情油然而生。他在心中默赞:竹子真了不得!

知识加油站

竹材的力学特征:竹材具有刚度好、强度大等优良的力学性质,是一种良好的工程结构材料。它的静弯曲强度、抗拉强度、弹性模量及硬度等数值比一般木材高约 2 倍。

森林的故事·竹子篇
The Story of the Forest · Bamboo Chapter

变身家用的竹

闲来无事，伯伯和小金果围着茶几，坐着喝茶，聊到竹子在人类生活中的作用。

小金果滔滔不绝："我们的生活可离不开竹子，比如夏季睡的竹凉席、竹床、竹垫，吃饭用的竹筷，剔牙用的牙签，我们坐的竹摇椅，奶奶买菜用的竹篮子，爸爸用的钓鱼竿，环卫工人用的竹扫帚，我房间里的竹书架，还有……"

伯伯点着头，用手敲了敲茶几，神秘地说："你看看这是什么材料做的？"

小金果瞪大了眼睛，左摸摸、右看看、上敲敲、下抬抬，还是看不出究竟，调皮地说："您又没教过我木材的知识，我怎么知道啊。反正是一种木头。"

"哈哈，这也是竹子的！"

小金果实在无法想通竹子和眼前这个光滑美观、色泽自然的茶几有什么关系。他将信将疑地说："不可能吧？这和我房间的竹书架完全不一样，我也根本看不到竹子的痕迹。"

"这是现代竹家具。传统的竹家具大都是直接用竹子手工扎

听听竹子的故事

制而成的，结构不够紧实严密，而且因为加工过程粗糙，产品容易生虫、发霉，还会受环境影响开裂变形。"

听伯伯这么一说，想想自己房间竹书架的模样，小金果认可地点了点头。

森林的故事·竹子篇
The Story of the Forest · Bamboo Chapter

"为了克服这些缺点,发挥竹子的功效,现在人们就把竹子做成板材再制作家具。这个茶几就是用竹板材制作的。"

伯伯起身领着小金果又欣赏了屋里刚添置的另几件竹板材家具,介绍说:"竹板材可以任意使用,实现家具时尚与环保、高贵雅致与舒适的完美结合,而且不积尘、不结露、易清洁,免去虫蛀之扰。"

"真是自然与科技的完美结合啊!"小金果啧啧赞叹。

知识加油站

竹材人造板:竹材人造板是以竹材为主要原料,经过一系列物理化学处理和机械加工,在一定的温度和压力下,经胶合压制而成的板材和方材。它消除了竹材的各向异性、材质不均和易干裂等缺点,具有卓越的物理学性能。

变形柔丝的竹

参观完竹子博物馆，伯伯和小金果走进了出口处的纪念品商店。店里陈列着琳琅满目的竹制品，除了竹扇、竹碗、竹杯、竹子制作的乐器玩具，居然还有毛巾、袜子、服饰和床品。这些物品一下子吸引了小金果。

"竹衣？"小金果眨巴着眼睛看向伯伯，"竹子真的可以做成衣服吗？"

伯伯笑道："嗯。从竹子中提取出竹纤维，像棉、麻、毛、丝一样，也可以纺纱织布做衣服啦。"

小金果愣住，伸手去摸摸竹纤维毛巾、竹纤维袜子、竹纤维服饰，感觉到一阵柔软滑爽，很是清凉。但他仍觉迷惑不解，问道："平时大家都喜欢穿棉麻的衣服，那竹衣它又有什么吸引人的优点呢？"

"这要从竹纤维的微观结构来解答。在扫描电子显微镜下观察，可以看到竹纤维具有细长的孔洞和表面沟槽，这种多孔结构让它拥有优良的吸湿性和放湿性，从而自动调节人体湿度，实现冬暖夏凉的功效；竹子里还有一种独特的物质，具有天然的抑菌、

听听竹子的故事

森林的故事·竹子篇
The Story of the Forest · Bamboo Chapter

防螨、防臭、防虫功能。这都是它优于棉、麻纤维制品的特点。"

伯伯拉起小金果的手："走,伯伯带你去选一件,让你和竹子来一个更亲密的接触!"

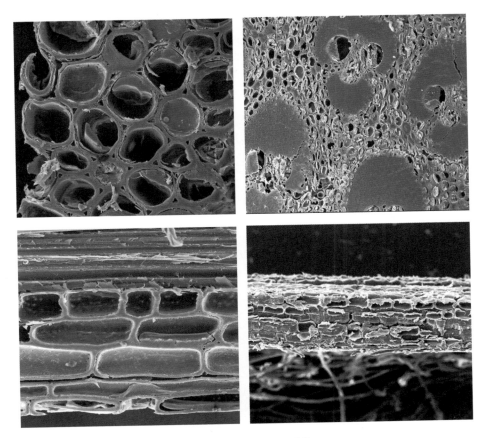

竹纤维的电镜显微结构

知识加油站

竹纤维：竹纤维是指从自然生长的竹子中加工提取出来的纤维素纤维。竹纤维细长空洞和表面沟槽的结构使其毛细管效应增强，从而具备优良的吸湿性和放湿性，成为纺织加工的首选。纺织用竹纤维按加工方法不同，分为竹原纤维、竹浆纤维和竹碳纤维三大类。

森林的故事·竹子篇
The Story of the Forest · Bamboo Chapter

可制良药的竹

初春时节,伯伯从竹林带回几棵翠竹,并在院子里支起了烤炉。看到这情景,小金果心中暗喜,追着伯伯的身影问:"伯伯,我们这是要做竹筒饭吗?"

伯伯看着小金果馋虫拱动的样子,笑着说:我们今天不做美食,而是要制一味良药——鲜竹沥。

制药?小金果好奇极了。

只见伯伯将竹竿锯成50厘米左右、比烤炉长度略长的竹筒,一劈两半,去除掉中间的竹节。然后将竹片横架在烤炉上,将炭

听听竹子的故事

火微微拨旺，并在竹片两侧正下方放置了几个玻璃杯。

炙烤十几分钟后，竹片开始吱吱作响，两端冒出小气泡，逐渐有液体渗出，恰好滴入底下的玻璃杯中。

小金果全神贯注地盯着伯伯的每一步操作。伯伯一边烤竹片，一边缓缓地说："我们现在收集在玻璃杯中的汁液就是竹沥，它具有清肺降火、化痰利窍的功效，是《中华本草》中记载的一剂良方，也是老百姓家常自制的备用药。"

伯伯取来一个杯子，倒入一些竹沥，再兑入3倍左右的矿泉水，递给小金果。小金果尝试着啜饮了一小口，一种清甜的味道，瞬间带着一丝清凉顺着喉咙滑下。

从根到梢全身都是宝。

没想到竹子还有药用价值。

看着小金果一副很享受的样子，伯伯接着说："从根到梢，竹子的很多部位都有药用价值。比如：淡竹叶能去热除燥、生津利尿；青秆竹的茎秆经去皮、刮丝、阴干而制成的竹茹性凉，能清热化痰、除烦止呕；竹卷心，也就是嫩竹叶卷而未张开的幼叶，具有清心除烦、消暑止渴的功效……"

"伯伯慢点，"小金果揉着脑袋，"我都快记不过来啦！"

伯伯笑了笑，像说相声似的又接了下去："还有竹黄，是竹子受伤后形成的分泌物经过凝固而成的块状物，具有清热解毒、定惊凝神的作用；竹衣，就是金竹秆内的衣膜，可治喉哑劳嗽；竹精是新竹管腔内的液汁，可治汗斑……"

小金果端着杯子，听得入神，此时他对竹子已经崇拜得五体投地了。对伯伯说道："难怪您常说'竹子全身皆是宝，入药治病把疾疗'。"

伯伯一边赞许地点了点头，一边将玻璃杯中的竹沥倒进玻璃罐，封口储存在冰箱里。

 知识加油站

竹沥：禾本科植物淡竹等的茎经火烤后所流出的液汁，具有清肺降火、滑痰利窍的功效。

净化空气的竹

小金果在伯伯的车上发现了一个有趣的小包。米白色布袋，外形像香囊，摸着沙沙作响，里面似乎装着许多珍珠小颗粒。

"伯伯，这是什么呀？"小金果好奇地问。

伯伯摩挲着小包，说："这是竹炭包，里面装着竹炭，是用来吸附车子里的异味、净化空气的。"

小金果拿起来闻了闻，隐隐的一阵竹香掠过鼻尖。"天呐，这简直和种了一棵竹子在车上一样神奇！这是怎么做到的啊？"

伯伯看着小金果贪婪呼吸的可爱样，笑着说："人们采用特殊工艺，将青翠苍绿的竹子变成为乌金发亮的竹炭。竹子的维管束组织中有无数肉眼看不见的细胞，细胞中的水分在烧炭过程中被抽光，留下炭化后的细胞壁，形成多孔质，让竹炭成了净化环境的最佳媒介。与木炭相比，竹炭具有更高的孔隙度，表面积是木炭的2～5倍，所以它的吸附能力远比木炭强。"

小金果捧着这个小巧却厉害的竹炭包，又问道："我知道用木材制成的活性炭可以净化废水，那竹炭它能做到吗？"

"当然喽！"伯伯十分骄傲地说，"竹炭的用处可大了。饮

听听竹子的故事

用水净化、土壤改良、废水处理、住宅调湿、除味消臭等，它都能实现，还可以制作竹炭燃料、竹炭纤维呢。够厉害吧？"

小金果瞪大了眼睛："简直太神通广大啦！不仅竹林可以净化空气，竹炭更是净化专家。"

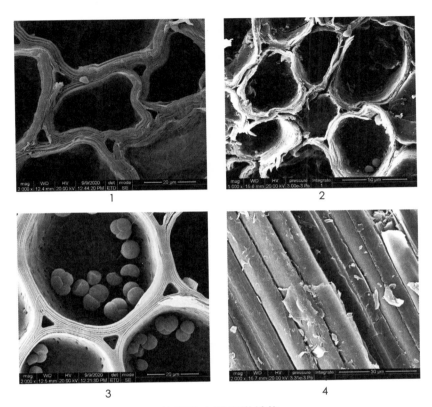

竹炭的电镜显微结构

1-3. 横截面　4. 纵切面

知识加油站

竹炭：竹炭是竹材热解得到的主要产品。通常选取4年以上成熟竹子，经高温无氧烧制而成。烧制竹炭的方法主要有两种：干馏热解法和土窑直接烧制法。

变幻妙音的竹

走出民乐团的演奏大厅,小金果还沉浸在悠扬婉转的音乐声中。他情不自禁地向伯伯描述自己的聆听感受:清亮的笛音、低沉的箫声、柔美的胡琴声……一切旋律都那么美妙。

"知道吗,这些民族乐器大多都是竹子做的。"伯伯说,"竹是八音之一。《史记》记载,远古时期,黄河流域生长着大量的竹子,于是当时的人们就开始选竹子为材料制作管乐器啦。"

听听竹子的故事

"人们为什么会想到用竹子来制作乐器呢?"

"大概是因为竹子具有圆润中空、管壁均匀、疏密适中的结构,使其振动性好、发音清脆,加之生长的普遍性使其取材方便,还有就是因为它便于加工。"

"那我回去,找根竹竿,像电视剧里一样做个竹笛吹吹。"小金果哈哈笑着说。

"从竹子到乐器,可不是简单的事。让一根生长在竹林中默默无闻的竹子变成发出悠扬乐曲的乐器,全凭匠人的一双巧手。需要取竹、阴干、去皮、烤竹、烫孔、校音、打磨……算下来,总共要有80多道工序呢。"

小金果吐了吐舌头,后悔自己刚刚冒冒失失的话。

"就拿做笛子来说吧。竹材要选用3年生以上的竹子,竹子的壁不能太薄,也不能太厚。太薄了声音发炸,轻飘飘的;太厚了声音发闷,所以壁厚一般选2毫米多一点的最合适。还有,也并不是竹子的所有部分都能用来做管乐器,能适用的也只是其中一段而已。"

"另外,竹笛打孔的位置也极为重要。毕竟竹子是天然生长,并非百分之百的标准圆柱形,所以吹孔通常要开在竹子的立面,即最突出的一面,这样吹奏起来最方便。此外,最为关键的便是调音。调音则归功于手艺人的灵敏听觉,在对笛孔进行微调时,一个不慎,多打磨下去一点点,笛子的音准可能就不对了,多少天的心血也就付之东流。"伯伯说。

不知不觉中,伯伯已领着小金果来到了后台。在这里,小金果见到了笛、箫、笙、竽等各式乐器,它们可全都是用竹子制成的啊!

 知识加油站

竹子与乐器:竹子与中国文化艺术有着不解之缘,它在音乐和美术方面的作用尤为突出,这与竹子结构的特殊性和分布的广泛性有很大的关联。中国乐器中用竹制成的非常之多,如笛、箫、笙、竽、竽等,不胜枚举。

爱 竹

情寄幽篁间

 森林的故事·竹子篇
The Story of the Forest · Bamboo Chapter

国画中的竹子

小金果和伯伯在北京故宫博物院参观，他被馆中一幅幅栩栩如生的中国画深深吸引住了。

眼前的这幅画：竹节坚韧遒劲；竹叶尖细有型；背景以淡墨勾勒出嶙峋的石块；清风中摇曳的劲竹与磐石相伴而生，傲气凛

听听竹子的故事

然。小金果觉得那棵竹子已跃然眼前,不由得大声吟诵:"咬定青山不放松,立根原在破岩中。"

伯伯赶紧拉拉小金果的袖子,示意他放低声音,然后说:"这正是板桥先生的《墨笔竹石图》,构图虽然简洁,但笔法十分稳健,将竹子的高洁素雅和坚韧不屈全都展现出来了。"

小金果问:"自然界中植物那么多,可这一路看下来,画竹子的却特别多,这是为什么呢?"

森林的故事·竹子篇
The Story of the Forest · Bamboo Chapter

伯伯回答:"中国传统文化喜借物喻志。古人把梅、兰、竹、菊称为'四君子',它们是写诗作画的常用题材。"

"竹子成为中国画题材的历史十分悠久。据文字记载,三国、两晋时期就有人以竹作画,南北朝隋唐五代时期得到发展,至唐朝则出现了直接以墨写竹的墨竹画法。到了元代,以写竹闻名的画家就有五六十人之多。明清之后,画竹的技法和形式更加丰富多样,竹画得到了极大的发展。"

边说边走,伯伯和小金果来到苏轼画的一幅竹图前。伯伯不无启发地指着它说:"画竹子时,心里要有一幅竹子的形象,才能一气呵成。这就叫——"

"胸有成竹!"小金果脱口而出。

 知识加油站

竹文化:我国是世界上竹类植物资源最丰富的国家。竹子作为我国重要的森林资源之一,其种类、面积和产量均占到全世界的三分之一左右,素有"竹子王国"之称。我国历史悠久,竹文化也丰富灿烂,是中华民族文化的重要组成部分,也是中华文化的精髓与瑰宝。

古诗中的竹子

《中国竹乡诗词大会》今晚将进行电视直播。这不,节目还没开始,小金果就已经拉着伯伯坐到了电视机前。

伴随着一段激昂的音乐,主持人宣布:"诗词比赛,飞花叙竹,开始——"小金果的心随即也被吊了起来,紧张地看着台上的选手进行对决。"竹外桃花三两枝,春江水暖鸭先知。"挑战者当仁不让领先一句。

"爆竹声中一岁除,春风送暖入屠苏。"擂主不慌不忙地接道。

"竹喧归浣女,莲动下渔舟""过江千尺浪,入竹万竿斜""六出飞花入户时,坐看青竹变琼枝""但对松与竹,如在山中时"……

中场广告间隙,小金果忍不住拉了拉伯伯的手,低声问道:"伯伯,有关竹子的古诗这么多啊!看来诗人一定很喜欢竹子!"

伯伯说:"是的。古代文人墨客钟爱竹子。因为它有秆、有枝、有节、有叶,中通外直,不畏严寒,一直以来都是诗画中托物言志的对象,竹文化也成为中华民族文化的重要组成部分。据统计,古人关于竹的诗、词、赋、曲等共有14000多首。"

小金果脸上露出了吃惊的表情。此时,激烈的飞花令比赛又

听听竹子的故事

森林的故事·竹子篇

继续进行了。

你一句:"郎骑竹马来,绕床弄青梅。"

我一句:"竹杖芒鞋轻胜马,谁怕?一蓑烟雨任平生。"

"可以食无肉,不可居无竹。"

"窗前亦有千竿竹,不识香痕渍也无?"

"……"

你来我往,小金果从一首首竹诗词中,感受到从古到今人们对竹子的丰富情怀,自己也心驰神往。

 知识加油站

竹子分布:竹子在中国的自然分布极广,南自海南岛,北至黄河流域,东起台湾,西迄西藏的雅鲁藏布江下游,约相当于北纬18°~35°和东经92°~122°。其中长江以南地区的竹种最多、生长最旺、面积最大。

物以载文的竹简

博物馆举办认识竹简的文化体验活动，小金果非常积极，拉着伯伯一早就来等候了。

讲解员在展台上缓缓打开一卷竹简，说："竹简，古代用来写字的竹片。简多用竹片制成，每片写字一列，写完一篇文章将所有竹片编联起来，称为'简牍'，是战国至魏晋时代的重要书写材料。"

小金果耳语伯伯："是不是东汉蔡伦改进造纸术以后，人们

听听竹子的故事

就不再用竹简啦?"伯伯笑而不答,示意注意听。

"纸张发明后,竹简牍又与纸张并行数百年,直至东晋末年,简牍的书写才结束。后来,到了唐朝还有史书用'罄竹难书'来形容一个人的罪大恶极,可见竹简使用的源远流长。"

讲解一结束,小金果就追问伯伯:"古人为什么会选竹子作为书写材料?"

伯伯解释道:"竹材从内至外分为竹黄、竹肌、竹青。内侧的竹黄面易着墨且墨不易磨灭;竹子生长快,一年成林,取材容易;同时竹片易加工,纹理紧密色差小,韧性强,不易损坏折断。这些优点促使先人选择竹子作为书写材料。"

最后的体验活动是让大家制作竹简:先是截取一段匀称的青竹竿,破竹削成宽窄均匀的竹片;接着烘烤干燥,再打磨穿线。一串竹简才算制作完成,其后就可以在上面写上自己喜欢的字句。

说着容易做起来难,小金果好一阵手忙脚乱,让他深深地感受到了古人制简的不易,也进一步了解到了竹子在承载中华文明中的巨大功劳。

知识加油站

竹壁:竹秆圆筒外壳称为竹壁,是竹秆表皮至髓腔的统称。竹壁从外向内分为竹青、竹肌(竹肉)、竹黄3层。

 森林的故事·竹子篇
The Story of the Forest · Bamboo Chapter

爆竹的前世今生

乡下的除夕真热闹！小金果和小伙伴们在院子里燃放烟花爆竹。

"爆竹声中一岁除，春风送暖入屠苏……"他们手拉着手转圈圈，高声吟唱着。

小金果转头问道："伯伯，古诗中的爆竹是不是指的鞭炮？"

听听竹子的故事

爆竹的前世今生
History of Firecracker

伯伯拿起一节鞭炮让小金果看构造，笑着说："鞭炮俗称爆竹，但古时候的爆竹可不是今天的鞭炮。最早叫爆竹，是因为燃竹而爆。据说从南北朝开始，在新年这一天，家家户户开门做的第一件事就是焚烧竹子。他们希望用竹子的爆裂声来吓唬驱逐怪兽'年'和瘟神恶鬼，以此保佑家人平安。后来人们发明了火药，制作鞭炮，用点燃炮仗来代替燃烧竹子，寓意除旧迎新、竹报平安。这民俗至今已有千百年的历史。"

小金果追问："烧竹竿为什么会发出噼噼啪啪的响声呢？"

森林的故事·竹子篇
The Story of the Forest · Bamboo Chapter

"竹竿有竹节，竹节使竹竿形成一个个密不透气的空间。青竹竿又富含水分，燃烧的高温将水变成水蒸气。持续受热，水蒸气不断膨胀，竹竿内压力不断增强，直至最终竹腔爆裂，从而发出巨响。"伯伯解释道。

"噢，我知道啦！只有烧青竹才会有爆破声，如果是干枯的竹子，那就不会响了。"小金果恍然大悟。

 知识加油站

爆竹溯源：爆竹在中国自起源至今已有2000多年的历史。在火药和纸张发明之前，古人用火烧竹子，使之爆裂发声，以驱逐瘟神。因竹子焚烧发出"噼噼啪啪"的响声，故称爆竹。

护 竹

大熊猫与竹

森林的故事·竹子篇
The Story of the Forest · Bamboo Chapter

选竹为食的大熊猫

打小开始,从故事书到动物园的雕塑,国宝大熊猫就给小金果留下了深刻的印象。它们有着大大的黑眼圈、长长的黑围脖、胖胖的肉身子、圆圆的大脸盘,还有肥肥的前爪子,抓着绿绿的翠竹枝。

听听竹子的故事

这一天,一走进动物园的熊猫馆,小金果首先看到的就是一只大熊猫,正直着上身、盘腿而坐,前爪抓着鲜嫩的竹子快速地塞进嘴巴里,像啃甘蔗一样咔嚓咔嚓,大腮帮也跟着一鼓一鼓的,憨萌可爱。

小金果转头问伯伯:"大熊猫为什么特别喜欢吃竹子呀?"

"是不是因为竹子长得快,一长一大片,大熊猫吃得很方便?"

伯伯正欣赏着憨态可掬的大熊猫,还未等及伯伯接话,小金果就自问自答道。

"说得对,但也不全对。"伯伯的回答总是那么吊人胃口。

"你知道吗?被誉为'活化石'的大熊猫在地球上已生存了800万年,它们的祖先可是吃肉的。"

"那后来怎么又变成吃竹的呢?"

"那是约1万8千年前,地球出现了大面积冰川等自然环境的剧烈变化,使得森林锐减,很多动植物随之灭亡。大熊猫缺少充足的可食用动物,通过逐渐改变食性,才存活至今的。"

"原来竹子是大熊猫的救命食物啊!难怪至今不离不弃呢。"小金果找到了答案。

伯伯继续说道:"把竹子作为主食的动物少,所以大熊猫的同食竞争者也很少,这使得它们有充足的食物。但它们也会偶尔逮着机会去掏只竹鼠、捅个鸟窝,恢复一下自己祖先的食肉本性,为自己开一次'小荤'。"

"那大熊猫看来是个'假和尚'呀!"小金果调侃道。

伯伯被小金果的幽默逗乐了。

 知识加油站

　　大熊猫与竹:大熊猫,是中国最古老的生物之一,被誉为古生物"活化石"。它主要生活在我国的高海拔地区,以竹为食。国外园林界把竹子和大熊猫确定为中国园林的标记。

食竹需要强大肠胃

伯伯和小金果饶有趣味地观察着大熊猫的一举一动。调皮的大熊猫吃饱了,就地拉下一个大粪球。小金果不自觉地以手掩鼻。

伯伯见了调侃说:"国宝就是不一样,粪便都带着清香味。"

小金果一脸木然。

伯伯笑着解释:"你想啊,大熊猫肠子短,竹子在里面停留很短时间,还没来得及消化完全就排出来了。没有发酵那不就带着竹子的清香味嘛。"

小金果极尽眼力观察那坨粪便,表面支离旁插,依稀看见长短不一的竹节纤维。他担心起来:竹渣子会不会刮伤了大熊猫的消化道?

小金果转头发现伯伯还在专心看,就没吱声。他看到:大熊猫把竹子咬成小节、塞满嘴巴,然后一起咀嚼,却不吐渣子。

小金果按捺不住,想询问伯伯。伯伯似乎也猜着了他的意思,不急不慢地说:"因为长期食竹,大熊猫的口腔已经进化出宽大的白齿,拥有强大的咀嚼肌,这些能帮助它们有效地磨碎粗糙的竹纤维。它们的消化系统也随之变得更强大,消化道黏膜肌的厚

听听竹子的故事

森林的故事·竹子篇
The Story of the Forest · Bamboo Chapter

度大约是人类的 8 倍，这都能帮助它们更好地消化竹子。不过，至今它们还不能完全消化竹子中的纤维和木质素。"

"所以它的粪便里还能看见清晰的竹叶与竹纤维。那尖锐的

竹枝会不会伤害它们自己呢？"小金果迫不及待地问。

"大熊猫的消化道里有着发达的黏液腺，能分泌大量黏液。取食时，它们将竹子包裹起来，增加了润滑度，避免了消化道受到伤害。排便时，这种黏液又包裹于粪便之外，减少了粗纤维的机械刺激，从而很好地保护自己。"伯伯回答道。

爱思考的小金果继续追问："竹子没有完全消化，那营养也不能全部吸收，可是大熊猫还是长得胖乎乎的？"

伯伯说："那是因为我们的国宝吃饭很努力。与'血统纯正'的植食性动物相比，尽管大熊猫对植物营养的利用率不足20%，但每天除了睡觉，它都会花十几个小时来吃20千克左右竹子，以补充身体的能量。剩下的时间则是在拉粑粑，每天要排便约40次，也很辛苦的呀。"

听完，小金果惊讶地张大了嘴巴，默默感叹：国宝生存真不易，大家应该都珍惜！

知识加油站

竹子的化学成分：竹茎主要由纤维素、半纤维素和木质素组成。通常，整竹由50%～70%的纤维素、30%的半纤维素和20%～25%的木质素组成。除此之外，还含有一定数量的蛋白质、氨基酸、脂类等。竹子的化学成分在不同的属种之间有一定差异。

森林的故事·竹子篇
The Story of the Forest · Bamboo Chapter

握竹特化长出六指

周末一到,小金果就又拉着伯伯来到了动物园熊猫馆,趴在围栏上目不转睛地观察大熊猫。

眼前的这只大熊猫正抱着竹子,啃得津津有味。只见它抓住竹秆,连同竹叶,快速地塞进嘴巴,然后滋滋地咀嚼。圆滚滚的大熊猫灵巧地抓竹食竹的样子,让小金果忍不住吐槽:"大熊猫就是一个标准的干饭人。"

忽然,小金果有了新发现——大熊猫有六根手指头?一、二、三、四、五、六!还真是六根手指!小金果兴奋地向伯伯宣告他的新发现:

"伯伯,大熊猫是'六指竹魔',它有六根手指头。你看,竹子下方有一根手指托着,上面还有五根手指抓着。它比我们人类的手多一指!"

伯伯被小金果的表情和动作逗得忍俊不禁,赞许地说:"没错。托竹的这一根指头,是由一节腕骨特化而形成的,学名为'桡侧笼骨'。由于大熊猫的爪子不像我们的手指那么灵活,为了方便握住竹子,久而久之,这节腕骨就变成了新'拇指'。"

听听竹子的故事

"原来是这样。"小金果若有所悟地说。

"但这根拇指又与人类的拇指不同。它是腕骨，里面没有指骨，也没有关节，所以不会弯曲，只是一个'伪六指'。它的作用在于帮助大熊猫稳稳地握住竹子，是大熊猫为了生存的需要而长出的'第六指'。"

小金果信服地点点头，感叹道："大熊猫可真聪明！为了做最佳'干饭人'，不断进化，形成新性状、生成新手指，造就食竹新本领。这应该也是'适者生存'的例证吧！"

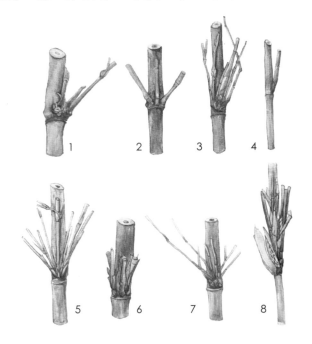

大熊猫食用的部分竹类品种
1. 八月竹　2. 实竹子　3. 短锥玉山竹　4. 箬叶竹
5. 少花箭竹　6. 巴山木竹　7. 缺苞箭竹　8. 冷箭竹

 知识加油站

　　大熊猫的食用竹：大熊猫可食竹60余种，主要有冷箭竹、八月竹、实竹子、筇竹、大叶筇竹、箬竹、少花箭竹、短锥玉山竹、峨热竹、巴山木竹、糙花箭竹、缺苞箭竹等。

竹子开花预示饥荒

成都大熊猫繁育研究基地，是我国开展大熊猫等濒危野生动物迁地保护工程的主要研究基地。在基地参观时，小金果听到路边仿石质音箱里传送来优美的音乐。

"竹子开花啰喂，咪咪躺在妈妈的怀里数星星。星星呀星星真美丽，明天的早餐在哪里……"听着歌曲，小金果高兴地说："我也会唱，我也会唱。"接着就不由自主地跟着哼唱了起来。

伯伯扭头问小金果："你知道这首《熊猫咪咪》歌曲背后的故事吗？"

小金果摇了摇头，说："我只知道咪咪是这只大熊猫的名字。"说罢他期待地望向伯伯。

伯伯沉思了一下，说："这是一首流行于20世纪80年代的歌曲，是为拯救濒临灭绝的国宝大熊猫而创作的。"

"熊猫灭绝？"小金果惊讶地问道。

"当时四川岷山北部的箭竹林出现大面积开花现象，使得生活在这一栖息地的大熊猫面临极为严重的粮食危机，随后就难觅熊猫踪影。为此，全国还发动了为大熊猫募捐的活动。"

听听竹子的故事

森林的故事・竹子篇
The Story of the Forest · Bamboo Chapter

竹子开花预示饥荒
Bamboo Flowering and Famine

"竹子开花、粮食危机，这两者有什么联系？竹子开花为什么会导致熊猫没饭吃呢？"小金果疑惑地问道。

"因为竹子整片开花，通常会导致竹林成片死亡。竹林没了，大熊猫就会断粮。"

"开花、结实、种子发芽，应该就是长出新竹呀，怎么会变成竹子死亡啦？"小金果迷惑了。

"和普通植物不同，竹子通常一生就开一次花。在生长前期，竹子的营养生长占优势。随后，生长优势逐渐转向生殖生长，直至最后开花结实。它们开花结实，需要消耗掉贮存在根、茎、叶中的大量有机养料。当这些营养器官中的养料被消耗完时，竹子就会相继枯死。"伯伯耐心细致地解释道。

"原来竹子开花反倒变成了竹子生长的终点，从而导致大熊猫闹起了饥荒。"小金果恍然大悟。

"所以，我们也要做好竹子资源的保护和竹子经营的可持续发展！"伯伯拍拍小金果，小金果坚定地点了点头。

 知识加油站

　　植物的开花习性：植物的开花习性可分为两类：一类是一生只开花一次的植物，如稻、麦、竹子等；另一类是多次开花植物，如苹果、梨等。

森林的故事·竹子篇
The Story of the Forest · Bamboo Chapter

护竹创造美好未来

伯伯外出考察已经有两个星期了。小金果很是想念伯伯。

这一天放学回家，小金果抬眼看见书房电脑前坐着一个人，知道是伯伯回来了。

小金果飞奔过去，泥鳅似的一下子钻到伯伯与书桌之间，并麻利地坐到了伯伯的腿上，一把搂住了伯伯的脖子。他太好奇伯伯的这次科考之旅，忙不迭地问道："伯伯伯伯，您看到野生大熊猫了吗？"

伯伯看着小金果迫不及待的神情，指着桌子上的相机笑着说："你和伯伯一起来整理这次科考的照片，怎么样？"

"好耶！这样我也可以通过图片来追寻您这次大熊猫与竹子的神秘科考之旅了！"小金果兴奋地回答。

将所有照片从相机导入电脑，并按时间地点进行归类，小金果协助伯伯做得有条不紊。浏览着一张张竹子的图片，伯伯介绍说："这些都是国宝大熊猫最喜欢的美食种类。"

"我知道。我国已知的竹子种类有600多种，大熊猫吃的有五六十种，但它喜欢的也只有20多种。我们的国宝是个挑食者。"

听听竹子的故事

伯伯被小金果的比喻逗笑了。这时屏幕上显示的大熊猫趴在树枝上的照片深深吸引了小金果。这棵树看起来有十几米高,大熊猫所在的位置距离地面至少也有七八米高,底下是茂密的竹丛。看着大熊猫那圆滚滚的身子,再看看树枝的顶端,小金果真是为国宝捏了把汗,担心它随时会压断树枝掉下来。

看着小金果睁大眼睛的紧张表情,伯伯安慰他说:"大熊猫是爬树高手。这只大熊猫饱餐后就爬到了树上,正美美地享受阳光浴呢。它是我们在秦岭佛坪三官庙保护区的密林深处拍到的。"

"三官庙,是不是早两天报道有5只大熊猫举行'比武招亲'的那个地方?"

"是的。"伯伯点了点头,赞许地说:"看来我们的小金果还搜集了不少的熊猫'情报'呀!你还掌握了什么?"

"我还知道,大熊猫的受威胁等级已经由'濒危'下调为'易危'啦。"小金果得意地说。

"大熊猫保护等级下降,反映了我国在生物多样性保护和生态修复方面取得了重大成就。这次开展的大熊猫野外栖息地竹子种质资源调查与可持续发展的研究课题,也是为了更好地开展竹子的可持续经营,同时更好地保护我们的国宝。"

"嗯,我知道。我们要经营好竹子,保护好国宝。"

"是呀。只有加强生态建设,做好森林资源的可持续发展,

森林的故事·竹子篇
The Story of the Forest · Bamboo Chapter

才能让大熊猫拥有一座更和谐的家园，一份更绿色的食品，一个更美好的未来。"

伯伯将手掌放到小金果的肩膀上，用力地压了压，仿佛是将这样的重任传递给了年轻的一代。

 知识加油站

　　森林可持续经营：森林可持续经营是指通过现实和潜在的森林生态系统的科学管理、合理经营，维持森林生态系统的健康与活力，维护生物多样性及其生态过程，以满足社会经济发展过程中对森林产品及其环境服务功能的需求，保障和促进人口、资源、环境与社会经济的持续协调发展。

参考文献

易同培. 四川竹类植物志 [M]. 北京：中国林业出版社，1997.

张齐生. 中国竹材工业化利用 [M]. 北京：中国林业出版社，1995.

江泽慧. 世界竹藤 [M]. 北京：中国林业出版社，2008.

康喜信，胡永红（上海植物园），等. 上海竹种图志 [M]. 上海：上海交通大学出版社，2011.

周芳纯. 竹林培育学 [M]. 北京：中国林业出版社，1998.

中国科学院中国植物志编辑委员会. 中国植物志 [M]. 北京：科学出版社，1991.

温太辉. 中国竹类彩色图鉴 [M]. 台北：淑馨出版社，1993.

周芳纯. 竹林培育学 [M]. 北京：中国林业出版社，1998.

彭镇华，江泽慧. 绿竹神气 [M]. 北京：中国林业出版社，2006.

李承彪. 大熊猫主食竹研究 [M]. 贵阳：贵州科技出版社，1997.

易同培，史军义，等. 中国竹类图志 [M]. 北京：科学出版社，2008.

周芳纯，胡德玉. 中国竹诗词选集 [M]. 南京：江苏古籍出版社，2001.

邹惠渝. 邵武竹类 [M]. 上海：上海科学技术文献出版社，1989.

陈守良，贾良智. 中国竹谱 [M]. 北京：科学出版社，1988.

史正军，杨静，杨海艳. 大型丛生竹材应用基础性能研究 [M]. 北京：科学出版社，2018.

王三毛. 古代竹文化研究 [M]. 北京：北京联合出版公司，2017.

后 记

竹子是中华文明的重要符号，中国是世界上竹林面积最大的国家。作为原产中国的一类植物，竹子不仅以其高风亮节、悠然淡雅的品质被列于"花中四君子"和"岁寒三友"，而且与大熊猫合并列为中国标记。从2018年4月我们动议编写这本有关竹子的科普图书开始，到2022年这个春笋萌发的季节全稿完成，创作历时整4年。捧出书稿，我们由衷地感到高兴。

该书由长期从事林业科研和教育的老师创作，以竹子为选题，讲解与竹子有关的科学知识，是一本渗透传统文化精髓和现代科学研究技术的本土原创科普读物。主创人员从知识点的钻研、趣味点的挖掘、故事的写作打磨到呈现画面的构图创作，最终完成了这样一本中英对照、图文并茂、音频和视频增色的科普读物。这是一本凝练传统文化与科学知识、汇集纸质媒体与现代传媒手段为一体的科普读物。

这本科普书的创作与编写，凝聚了专家学者、老师学生的智慧与才艺。编写过程中，周吉林老师对故事体系进行了指导；丁雨龙、张春霞老师和中国大熊猫保护研究中心杨建对科学知识点进行了审核；刘冬冰、张武军等老师参与了部分文字的编辑工作；包心悦、傅珺旻、何玲、陆彤等同学参与了素材的收集与编撰工作；卫欣、贾文婷、耿植荣、赵亚洲、宋昱萱老师参与了情境图与科学图的创作与指导工作；蔡佳沁、李悦、张瑾钰、梁家昱、吴璇、舒画等同学参与了情境图与科学图的绘画工作，南京林业大学现代分析测试中心提供了精美的电镜图片。在此，对所有为此书出版做出贡献的各位老师和同学表示最诚挚的感谢！限于编者的知识体系与能力水平，书中可能存在不足，敬请读者不吝赐教。

编者
2022 年 11 月

The Story of the Forest

Bamboo Chapter

Yang Qing
Editor in Chief

China Forestry Publishing House

图书在版编目（CIP）数据

森林的故事.竹子篇:汉、英/杨青主编.--北京：中国林业出版社，2022.12

ISBN 978-7-5219-2072-7

Ⅰ.①森… Ⅱ.①杨… Ⅲ.①森林—儿童读物—汉、英②竹—儿童读物—汉、英 Ⅳ.① S7-49

中国国家版本馆 CIP 数据核字（2023）第 001000 号

森林的故事·竹子篇
The Story of the Forest · Bamboo Chapter

数字资源信息

策划编辑：于界芬　吴　卉
责任编辑：吴　卉　于界芬　倪禾田
出版发行：中国林业出版社
（100009，北京市西城区刘海胡同7号，电话83143552）
电子邮箱：books@theways.cn
网址：https://www.cfph.net
印刷：河北京平诚乾印刷有限公司
版次：2022 年 12 月 第 1 版
印次：2022 年 12 月 第 1 次印刷
开本：787mm×1092mm 1/16
印张：17
字数：147 千字
定价：88.00 元（中英文版共 2 册）

Series Editor: Cao Fuliang

Editor in Chief: Yang Qing

Deputy Editor in Chief: Cao Lin, Wang Tao, Jia Wenting

Committee members and division of labor

Overall planning: Zhou Jilin, Yang Qing

Story Creation: Yang Qing, Bao Xinyue, Li Yangnan, Cao Manman

Illustration creation: Geng Zhirong, Jia Wenting, Zhao Yazhou, Yang Qing, Song Yuxuan

English translation: Wang Tao, Cao Lin

Scientific Advisor: Ding Yulong, Zhang Chunxia, Yang Jian

Artist Consultant: Wei Xin

Preface

Dear young friends, bamboo is an amazing plant. In the spring of 1982, I graduated from university and worked in the Bamboo Research Institute in this university. At that time, I often worked with Mr. Zhou Fangchun in bamboo producing areas such as Yixing in Jiangsu Province, Mogan Mountain in Zhejiang Province, Yifeng in Jiangxi Province, and Guilin in Guangxi Zhuang Autonomous Region, where we often stayed for a month or two, to carry out scientific research work on bamboo, soil sampling, bamboo resources investigation, yield estimation, and so on. I had been doing research on bamboo for more than four years. Through close contact, I had gained a deeper understanding of bamboo. With the continuous accumulation and enrichment of relevant knowledge, I am more aware that the economic, ecological and cultural values of bamboo are unparalleled in the plant world. Gradually, I began to have an idea of writing a bamboo science book, through which you can learn more about bamboo and improve your scientific literacy for the sake of your growth, the nation as well as the whole planet.

Bamboo has a big family. There are more than 1400 varieties distributed in tropical, subtropical and temperate zones throughout the world, roughly more than 600 of which grow in China, the hometown of bamboo. Archaeological findings show that bamboo plays an important role in the lives of our ancestors in building bridges and houses, making furniture and farm implements and even turning bamboo shoots into delicious food. Especially in the Shang and Zhou

Dynasties, our smart ancestors made slips from bamboo instead of oracle bones to record words.

Today, when the earth is getting warmer and the environmental problems are prominent, bamboo is a good partner of mankind as its capacity of soil conservation, CO_2 absorption, carbon fixation and oxygen release is much higher than that of other plants. After bamboo planting, it can be thinned in 3 to 5 years and has very good economic benefits as industrial materials. Bamboo has a graceful figure, delicate green leaves, straight and modest branches, and it

can grow in all kinds of soils, which are similar to many virtues of our nation. Therefore, bamboo has also become the object of praise in Chinese culture.

Looking at bamboo from the perspective of botany, we will find that it has many fascinating characteristics that are hard to ponder. Answers to the questions such as "Is bamboo grass or tree?", "Why does bamboo grow so fast?", "Why does bamboo grow taller but thicker?", "Why bamboo can grow into a bamboo forest?", and "Why bamboo blossoms and then withers only once in its life" will be unraveled when you finish reading this book!

Bamboo research has a bright prospect! In 1997, the International Network for Bamboo and Rattan (INBAR) was established in Beijing and it is the first intergovernmental international organization with its headquarters in China with 48 member states at present. It not only marks the important influence of China in bamboo and rattan research, but also shows the broad prospect of bamboo and rattan culture research. I hope you can learn some knowledge about bamboo from this book, love it, have more ideas and take actions to explore the mystery of plants in a bid to make the land greener and the mountains and rivers more beautiful with your efforts.

Cao Fuliang
October, 2022
Nanjing

Foreword

Dear young friends, this is a popular science book about bamboo both in Chinese and English version presented in the dialogue between Uncle Tony and Little Ginkgo and supplemented with vivid and lively hand-drawn illustrations and audio reading. It introduces us the knowledge about bamboo with fresh and lively style and interesting expressions. Moreover, it is a light and pleasant reading and can enrich your mind on reflection. You can reap a lot from reading the book.

Bamboo is a familiar plant to us. Botanists say it is a kind of highly lignified herb, environmentalists say it is the most friendly plant to the earth and men of letters say it embodies virtues such as faithfulness, fortitude, straightness, modesty and integrity and give it many beautiful names such as "Huang" "Yun" and "Baojiejun".

The book consists of six parts.

In Part One, we can know why bamboo has a good figure, why bamboo shoots grow so fast after spring rain, how bamboo rhizomes grow underground and many secrets of bamboo growth.

In Part Two, we can appreciate the "-est" of bamboos such as the tallest, the shortest, vine-like climbing bamboo, *Phyllostachys bambusoides* 'Lacrimadeae', *Phyllostachys aureosulcata*, etc. as if walking into the Bamboo Expo.

In Part Three, we can learn some scientific knowledge about bamboo cultivation and management, understand the application of cloning technology in bamboo seedling culture and even learn how to dig bamboo shoots.

In Part Four, we know that bamboo is not only the raw material for building materials, furnitures, medicine and musical instruments, and through modern technology, bamboo can be transformed into bamboo charcoal to purify the air, and bamboo fiber can be made into textiles. The study of bamboo has attracted the attention of scientists from many countries.

In Part Five, we may know the special place of bamboo in Chinese culture. Bamboo, together with plum, orchid, and chrysanthemum, is claimed to be "the four gentlemen", which not only appears in calligraphy and painting, but is also the inheritor of culture and the carrier of folk customs.

In Part Six, we will read the story of giant pandas and bamboo, and know it is not easy for giant pandas to choose bamboo as their food in the process of evolution. We should make our efforts to protect bamboo for giant pandas as well as for a better life.

We have all read Zheng Xie's poems, "Upright stands the bamboo amid green mountains steep; its toothlike root in broken rock is planted deep. It's strong and firm though struck and beaten without rest, careless of the wind from north or south, east or west." The tenacity and perseverance of bamboo have aroused our admiration and respect. Now let's read *The Story of the Forest · Bamboo Chapter* happily and bear all the scientific knowledge about bamboo in our mind, which may help us grow. Dear children, it will be our common good fortune if some of you become professionals in a certain aspect related to bamboo in the future after reading the book.

Yang Qing
October, 2022

Contents

Preface IV

Foreword VII

Physiology — Fascinating Bamboo 1

1. Neither Grass nor Wood 2
2. Blossom and Fruit Bearing 5
3. Bamboo that Grows Tall but Not Thick 8
4. Vital Bamboo Rhizome 11
5. Fast Growing Bamboo Shoots after Spring Rain 14
6. Magical Phenomenon — Bamboo Spitting Out Water 17
7. Siblings or the Same Generation 20

Evergreen Bamboo — Variety 23

8. The World Record among Bamboos 24
9. The Three Bamboo Families 27
10. The Vine-like Climbing Bamboo 31
11. *Bambusa ventricosa* with Fat Waist and Round Belly 35
12. Bamboo Inlaid with Gold and Jade 38

13. *Bambusa multiplex*   41

14. *Phyllostachys bambusoides* 'Lacrimadeae' Sprinkled with Thousands of Tears   44

Bamboo Cultivation — No bamboo is vulgar   47

15. Scientific Ways of Bamboo Breeding   48

16. Scientific Cloning of Small Bamboo Seedlings   51

17. Skills in Digging Bamboo Shoots   54

18. *Phyllostachys edulis* Tabbing and Filing   58

19. Cutting Bamboo Tips to Resist Snow Disaster   61

Utilization — No Residence, No Bamboo   63

20. One Pole as a Boat   64

21. Bamboo for Landscaping   68

22. Bamboo that Is as Strong as Steel and Iron   71

23. Domestic Use   74

24. Bamboo that Transforms into Soft Silk   77

25. Bamboo that Can Be Used to Make Good Medicine   80

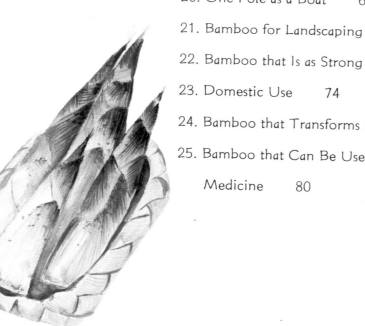

26. Air Purification   83

27. Bamboo that Transforms Air into Melody   86

Culture — Poetic Bamboo   89

28. Bamboo in Traditional Chinese Painting   90

29. Bamboos in Ancient Poems   93

30. Bamboo Slips for Writing   96

31. History of Firecracker   98

Symbol of China — Bamboo and Giant Panda   101

32. A Bamboo Eater   102

33. A Strong Stomach   105

34. Six Fingers   108

35. Bamboo Flowering and Famine   111

36. Protecting Bamboos to Create a Better Future   114

References   118

Epilogue   119

Chapter I Physiology
Fascinating Bamboo

森林的故事·竹子篇
The Story of the Forest · Bamboo Chapter

Neither Grass nor Wood

During the summer holiday, Little Ginkgo followed Uncle Tony to Jiangxi Province for inspection.

The car driven on the winding mountain road, with green bamboo forest on one side and rolling rice paddy fields on the other, the scenery was very pleasant.

While stopping to have a rest and looking at the fascinating scenery, Tony told Little Ginkgo, "Jiangxi Province is a major production area of bamboo and a hometown of rice. Do you know that bamboo and rice are close relatives?"

"Bamboo is tall and straight like a tree, how can it be a relative of the weak 'straw'? Does bamboo belong to herb plant?" Little Ginkgo asked suspiciously.

Tony replied, "Herbal plants are annuals. Their stems are short and soft. However, bamboo is perennial. Its stem is strong and tall, with a high degree of lignification. Therefore, bamboo is not grass."

"Do you mean bamboo is a kind of tree?" asked Little Ginkgo while looking at Tony.

Tony explained patiently, "No. Trees not only have xylem and phloem, but also have cambium, so that they can grow thicker every year, while bamboo can't since it does not have this layer."

Little Ginkgo tilted his head, looking a little puzzled.

Tony continued, "Bamboo is a perennial evergreen plant. There are trees, shrubs, vines, and herb-like ones. It is really not easy to classify bamboo."

"That is to say, bamboo is neither grass nor wood." Little Ginkgo tentatively came to his conclusion.

Hear the story of bamboo

Tony said approvingly, "Yes, bamboo is a very special kind of plant that is neither grass nor wood."

"Please tell me more about this special plant, Uncle Tony." pleaded Little Ginkgo enthusiastically.

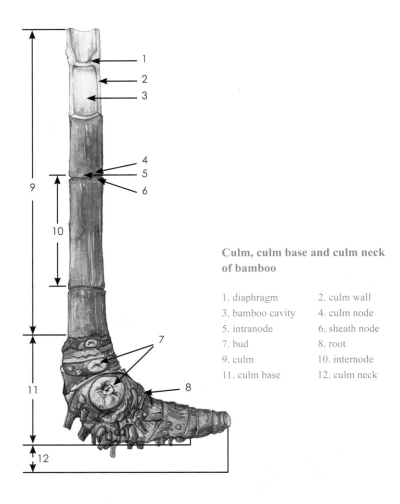

Culm, culm base and culm neck of bamboo

1. diaphragm
2. culm wall
3. bamboo cavity
4. culm node
5. intranode
6. sheath node
7. bud
8. root
9. culm
10. internode
11. culm base
12. culm neck

 Knowledge Point

Degree of Lignification: Lignin is one of the components that make up the xylem cell walls of plants. It keeps the xylem extremely rigid to bear the weight of the entire plant. Nearly half of the bamboo cell walls contain a large amount of lignin, so that with a high degree of lignification, the bamboo culm is hard and rigid.

Blossom and Fruit Bearing

Spring comes and flowers bloom; grass's lush and orioles fly. In the spring after COVID-19 pandemic, Little Ginkgo loved to go out for a walk with Uncle Tony in the park, looking for the beautiful traces of various plants.

There were bright yellow winter jasmines, pink early cherry blossoms, snow-white pear blossoms, bright red peach blossoms, purple tulips, bright red peonies... Little Ginkgo recorded all these colorful flowers in his *Plant Observation Diary*. Through careful observation, he even recorded the flowers of pine trees, Chinese hollies, and papaya trees that he had not paid attention to before. However, Little Ginkgo wondered why he never saw a bamboo flower though he had searched everywhere so carefully.

Passing by a bamboo forest today, a series of questions came to his mind again. Little Ginkgo couldn't help asking, "Uncle Tony, does bamboo blossom?"

"Yes. Bamboo is a seed plant. It will blossom and bear fruit."

"But why have I never seen a bamboo flower?"

"That's because bamboo has a very long flowering periodicity, which takes more than ten years, decades or even more than a hundred years to blossom. For example, it takes 120 years for *Phyllostachys bambusoides* to blossom. Therefore, bamboo flowers are not common."

"What does the bamboo flower look like?"

"The flower of bamboo is similar to that of rice because both bamboo and rice belong to Gramineae. Their flowers are called glume. *Phyllostachys edulis*, which is commonly seen, has yellow-white florets. There are two bracts under

Hear the story of bamboo

each floret, called lemma and palea. Pedals have been evolved into lodicule, so bamboo flowers have no bright colors. The low-hanging spikelet, which is attached to the side branches of bamboo, is composed of one or more florets, inner and outer glumes as well as rachilla." Tony said while showing the picture in his mobile phone to Little Ginkgo.

"The slender filaments hang with plump anthers, much like a string of wind chimes. This structure might help wind to disperse pollen, right? What does its

seed look like?" asked Little Ginkgo.

"The fruit of Phyllstachys edulis is similar to that of rice and wheat, which belongs to caryopsis. As caryopsis contains a large proportion of starch, the fruit of bamboo is called 'bamboo rice', which can also be used as food."

"Bamboo fruits are very valuable because bamboos rarely bloom. I want to try some if there is a chance." said Little Ginkgo, picturing a bowl of delicious bamboo rice in his mind.

Bamboo flowers and fruits of Pleioblastus yixingensis

 Knowledge Point

Seed Plant: Plants are classified into spore plants and seed plants. Seed plants can blossom and bear fruits.

The Story of the Forest · Bamboo Chapter

Bamboo that Grows Tall but Not Thick

Little Ginkgo has been to many bamboo towns with Uncle Tony, such as Anji in Zhejiang Province, Yifeng in Jiangxi Province, Chishui in Guizhou Province, Guangning in Guangdong Province, Shunchang in Fujian Province and so on. In his impression, countless bamboos are almost identical. With several joints, they are straight and slender, tall and straight. Green bamboo leaves are endearing.

This time in Yunnan Province, Little Ginkgo was deeply attracted by the row of tall bamboos on the roadside, and couldn't help shouting, "Look! Uncle Tony, how thick those bamboos are!"

Tony smiled and said, "Yes! They are thicker than your waist."

Little Ginkgo wrapped his arms around a bamboo culm and guessed, "It must have grown for many years. Is it 'ancient wood' or 'ancient bamboo' among bamboos?"

"This is *Dendrocalamus sinicus*. It's not ancient. This one is less than two years old!"

"Can it grow this thick within two years?" Little Ginkgo's jaw dropped. It occurred to him that the bamboo growing in their family backyard seemed to have never changed for years. It had maintained a slender figure since bamboo shoots were unearthed in the spring.

Knowing that Little Ginkgo was still confused, Tony smiled mysteriously and said, "The thickness of bamboo depends on the bamboo shoot. Come on,

Hear the story of bamboo

let's go to explore the mystery of bamboo's growth."

Tony asked Little Ginkgo to squat next to a bamboo stake, and said, "Look closer, a bamboo wall has three layers: the outer, middle and inner layers. The outermost green part is bamboo bark, also called bamboo green, the innermost light-yellow part is called tabasheer, and the middle part in between is bamboo flesh/meat made up of vascular bundle and fundamental tissues." Tony tore off a piece of translucent membrane from the bamboo's inner wall, handed it to Little Ginkgo and then said, "This can be used as a flute membrane."

Little Ginkgo was keen to get an answer, and interjected, "Why are bamboos of one variety, regardless of age, of almost the same thickness?"

Though seeing through Little Ginkgo's mind, Tony answered slowly, "The key lies in the vascular bundle. For a tree, the cambium between the phloem and

xylem of the vascular bundle has a meristem function. It differentiates outward to continuously form phloem, and differentiates inward to continuously form xylem. With the differentiation, the trunk becomes larger and thicker. However, for bamboo, there is no cambium between phloem and xylem of its vascular bundle. In the shoot stage, bamboo mainly depends on the growth of somatic cells to grow thick. After it breaks through the ground, the growth rate of somatic cells slows down, so it is difficult for bamboo to increase its thickness."

"That's how it is!" Little Ginkgo exclaimed, "A minute difference makes things completely different! Nature is amazing!"

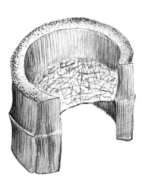
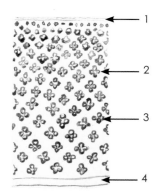

Distribution of vascular bundles in bamboo culm wall of *Phyllostachys edulis*

Structure of cross section of bamboo culm wall
1. bamboo skin/surface   2. fundamental tissue
3. vascular bundle   4. extramedullary tissue

 Knowledge Point

The Vascular Bundle of Bamboo: Bamboo belongs to Gramineae, whose vascular bundle is arranged outside the pulp cavity and inside the epidermis of the bamboo culm, and is scattered in the parenchyma, which mainly plays the role of transporting nutrients and water. There is no cambium in the vascular bundle and new cells cannot be generated.

Vital Bamboo Rhizome

A few bamboos were planted at the corner of Uncle Tony's yard. After a spring rain, a surprised exclaim came from the neighbor outside the fence, "Why are there a group of bamboo shoots in my yard?"

Little Ginkgo heard the voice and ran out of the house. He saw several bamboo shoots sprout from the ground in the neighbor's yard. They were very cute. Tony also came over to see what was going on.

"Alas, it's my fault!" Tony patted the back of his head and said, "I'm terribly sorry for the inconvenience they brought to you!"

Little Ginkgo was confused. Why could it be Uncle Tony's fault?

Tony found a shovel and carefully dug the soil here. Little Ginkgo found a slender "stem" that stretched to the bamboo forest in Tony's yard. There were several bamboo shoots growing on the stem.

Tony pointed and said, "This is called bamboo rhizome. It is a rhizome of bamboo. There are buds and adventitious roots on it. These buds will grow into bamboo shoots or new bamboo rhizomes."

The bamboo rhizome is the organ of bamboo to transport and store water and nutrients, and it is also the reproductive organ. As long as the season, rainwater, and temperature are suitable, it will randomly expand in all directions underground, or suddenly a big bamboo shoot will emerge.

Little Ginkgo asked eagerly, "But why did you say it was your fault?"

"Think about it. We planted bamboos in our own yard, but we didn't control their growth so that bamboo rhizomes expanded at random and gave our

Hear the story of bamboo

neighbor a big surprise. When planting bamboos, people usually build a fence by using cement boards around the designated area, to prevent the bamboo rhizome from growing everywhere."

Little Ginkgo giggled, "I see. Bamboo rhizomes are really naughty!"

Why can bamboos grow everywhere?

With strong reproductive ability, bamboo rhizome can expand as long as conditions are suitable.

任意行走的竹鞭
Vital Bamboo Rhizome

Morphological characteristics of bamboo rhizome

 Knowledge Point

Bamboo Rhizome: Bamboo rhizome is a common name for the underground stem of bamboo. It has obvious node, on which there are roots and on the sides of which there are buds. It can germinate into new underground stems or develop into shoots, and then sprout from the earth to grow into bamboos. Bamboo rhizome can pierce through hard soil, even through cracks in rocks, brick walls, and concrete or can cross obstacles to continue growing.

 森林的故事·竹子篇
The Story of the Forest · Bamboo Chapter

Fast Growing Bamboo Shoots after Spring Rain

It had been raining intermittently for three consecutive days. The sky finally cleared up.

Little Ginkgo couldn't wait to play outside. He ran into the bamboo forest, walked on a narrow winding trail, and found many bamboo shoots. Some just stuck out their small heads, and some had already grown higher than him. All of them, tall or short, were wrapped in dark brown coats.

Seeing so many bamboo shoots, Little Ginkgo was greatly surprised, touching one after another. Tony smiled and asked him, "Are you wondering why there are so many bamboo shoots in the bamboo forest all of a sudden?" Little Ginkgo nodded approvingly.

Tony continued, "Some researchers have measured that a bamboo shoot can grow nearly 1 meter in height overnight."

"That is to say it can grow nearty 1 millimeter in height every minute!" Little Ginkgo calculated it carefully, and then asked, "Why does bamboo shoot grow so fast after the rain?"

Tony answered jokingly, "It has drunk enough water in the rain and gathered strength underground. As soon as the small bamboo shoot sprout from the earth, it will grow rapidly!"

Little Ginkgo seemed not satisfied with Tony's explanation, pouting his lips with his eyes fixed on his uncle.

Tony picked a bamboo shoot, dug it out, and then split it in half from top to

Hear the story of bamboo

bottom, asking Little Ginkgo to observe it carefully. Little Ginkgo found that the bamboo joints inside were tightly folded, like a compressed spring.

Tony cleared his throat and said, "Bamboos grow faster than trees because trees have only one apical growth point, while the multi-segmented bamboo has a meristem in each node. The bamboo shoot sprout from fertile soil. In warm and humid weather, each meristematic tissue of the bamboo shoot will continuously form new cells, and the distance between the adjacent bamboo joints will be rapidly elongated, so that the growth rate will increase exponentially."

Seeing Little Ginkgo's puzzled face, Tony further explained, "Let me compare the growth of plants to the construction of tall buildings. The growth of trees is like construction using traditional techniques, which can only be constructed from bottom to top, layer by layer. However, the growth of bamboo

is like construction using a modern frame mode. Bamboo joints are building frames, and each bamboo joint is a construction joint. If all floors are constructed at the same time, the bamboo building can be built in a very short time."

While explaining, Tony looked up at the top of the bamboo shoot that would pierce the sky. So did Little Ginkgo, murmuring, "Well, I'm going to grow as fast as the bamboo shoot, and grow taller than Uncle Tony!"

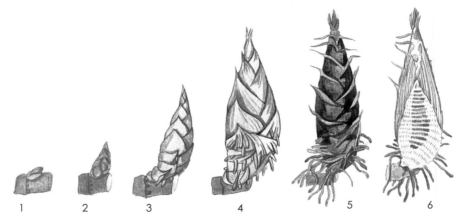

Bamboo shoots grocoth of *Phyllostachys edulis*
1. dormant lateral buds on the rhizome   2. germinating bamboo shoots
3. newly formed bamboo shoots   4. winter shoots before excavation/spearing out of the soil
5. spring shoots   6. longitudinal section of spring shoots

 Knowledge Point

Growth of Bamboos: During the growth stage of bamboo shoots in the soil, the apical meristem continuously performs cell division and differentiation to form nodes, internodes, internode septa, shoot sheaths, lateral buds and intermediary meristems. No new sections will be added afterward. The growth of bamboo shoots starts from the base. First, the bamboo shells grow, and then the intermediate meristem divides and grows node by node, which pushes the bamboo shoots to move up, sprout from the soil layer, and grow out of the ground.

Magical Phenomenon – Bamboo Spitting Out Water

It was a silent moonlit night with thin clouds and sparce stars. It was the first time for Little Ginkgo to conduct scientific research in the bamboo forest with Uncle Tony.

Suddenly, Little Ginkgo seemed to hear the sound of tick-tock raindrops. He looked up at the bright moon and asked Tony inexplicably, "Eek! It isn't raining! But why did I hear the raindrops?"

Tony smiled and said, "It's not raining, it's the sound of bamboo shoots spitting out water." Under the light of the flashlight, Tony squatted down with Little Ginkgo and observed it carefully. Little Ginkgo found that bright water droplets hung on almost every bamboo shoot. Those droplets were like shining pearls under the strong light of the flashlight, some of which were still rolling down. The roots of the bamboo shoots were wet.

Little Ginkgo asked curiously, "Are they dewdrops of the bamboo?"

Tony explained, "They are not dewdrops, they are water spat out by bamboo shoots, which is a unique phenomenon of bamboo shoots. They need a lot of nutrients, which are usually transported through water, for rapid growth. Therefore, it seems that a small water pump is installed in its root system, which absorbs a large quantity of water and nutrients in the soil and continuously delivers them to the bamboo shoots. The nutrients are quickly absorbed by the bamboo shoots, and the excess water will be spat out along the leaves, moistening the

Hear the story of bamboo

surface soil around the young bamboo, thus forming a small water cycle in the bamboo forest!"

Tony paused, then continued, "Bamboo spits out more water and forms larger water droplets especially at nights with high temperature and high humidity. The sound of countless water droplets dripping on the dry bamboo leaves on the ground is similar to the sound of dripping rain."

After hearing this, Little Ginkgo suddenly realized, "The sound of rain dripping in the bamboo forest at night was caused by bamboo shoots spitting out water. Haha!"

神奇的竹笋吐水
Magical Phenomenon - Bamboo Spitting Out Water

Bamboo shoots spitting water

 Knowledge Point

Bamboo Shoots Spit out Water: The phenomenon of bamboo shoots spitting water usually occurs at night when the temperature and humidity are high, and the water vapor in the air is nearly saturated. At this time, the evaporation of bamboo shoots is reduced, but the root system still substantially absorbs water, resulting in the phenomenon that the amounts of water inhalation is greater than that of water consumption, so the excess water will be discharged from the tip of bamboo shell leaves, forming water droplets, especially in hot humid summer.

森林的故事・竹子篇
The Story of the Forest・Bamboo Chapter

Siblings or the Same Generation

When spring breeze blows and spring rain falls, many bamboo shoots appear in the bamboo forest. Little Ginkgo and Uncle Tony went to the bamboo forest to dig bamboo shoots. Little Ginkgo touched them excitedly. The bamboo shoots had pointed and hard tips and resembled cute little babies.

Little Ginkgo asked, "Uncle Tony, are bamboo shoots the kids of bamboos? Are bamboo shoots supposed to be the seeds of bamboos?"

"Nope." replied Tony, while using a shovel to remove dead leaves and floating soil on the ground. Then an underground network was revealed, which was formed by the underground bamboo rhizome that spread vertically and horizontally.

"Bamboo shoots are the buds on the underground stems of bamboos, that is, the buds of bamboo rhizome. Bamboo shoots grow into bamboos." Pointing to a bamboo shoot on a bamboo rhizome, Tony continued, "Bamboo will blossom and bear fruit to breed the next generation, but the seeds produced by most bamboo blossoms are not fully developed. Reproductive seeds may be produced by only a few bamboo species."

"Oh, I see! The seeds after bamboo blossoms are the children of bamboos." Little Ginkgo said thoughtfully.

"Here comes the interesting thing!" Tony paused and said, "These new and upcoming bamboo shoots or bamboos are not strictly the 'next generation' in

Hear the story of bamboo

this bamboo forest. If we compare the bamboo forest to a tree, those newly grown bamboo shoots and bamboos are more like newly grown buds or twigs."

Little Ginkgo suddenly seemed to understand and said, "Oh, no matter whether those bamboos on the ground are old or young, if the underground stems (bamboo rhizome) are of the same age, they are siblings of the same root."

Why are the newly emerged bamboo shoot and the grown bamboo of the same generation?

As long as the bamboo shoot and the bamboo have the same bamboo rhizome, they are of the same generation.

Morphological characteristics of bamboo and bamboo shoots

 Knowledge Point

Bamboo Forest and Bamboo Tree: In botany, for a bamboo forest or clump connected underground, an underground stem is the main stem of a "bamboo tree", a bamboo culm is the branch of the "bamboo tree", and a bamboo shoot is the bud of the "bamboo tree". All of them form "a bamboo tree".

Chapter 2 Evergreen Bamboo

Variety

森林的故事・竹子篇
The Story of the Forest · Bamboo Chapter

The World Record among Bamboos

Uncle Tony and Little Ginkgo visited the Greenery Theme Park, where there were a wide variety of bamboos, including the elegant and stylish *Bambusa multiplex*, the grotesque-shaped *Phyllostachys edulis* 'Kikko-chiku', the charmingly naive *Bambusa ventricosa*, the golden yellow *Indosasa levigata*, the precious and peculiar *Phyllostachys edulis* 'Tao kiang', the elegant and beautiful *Phyllostachys bambusoides* 'lacrimadeae' and so on. What a variety!

Hear the story of bamboo

Seeing a lawn in front of him, Little Ginkgo delightedly cried out. He wanted to walk up and roll on it. While approaching, he noted that these grasses were different. They were lying prostrate, with linear-lanceolate leaves arranged in two rows and densely covered with hairs at nodes.

Seeing Little Ginkgo's astonished eyes, Tony took hold of a piece of grass and said, "This is not grass, but a kind of dwarf bamboo in the world which is no

more than a foot high, known as *Pleioblastus distichus*."

With great interest, Little Ginkgo blurted out, "If this is the smallest bamboo, then what is the largest?"

Tony looked as if he wouldn't be fazed by the question and replied, "Well, the largest bamboo in the world is the *Dendrocalamus sinicas*. It has the tallest and biggest culms which can reach up to a dozen floors in height, and up to an adult's thigh in thickness. It is also called as "the bamboo king", a rare and unique bamboo species in China, distributed in the southwest of Yunnan Province."

Little Ginkgo shook Tony's arm and said, "Uncle Tony, I have appreciated *Dendrocalamus sinicas* in Yunnan Province with you."

Tony said with certainty, "We will also take a look at various bamboos such as the thinnest and thickest bamboos, solid bamboos, those with longest and shortest joints and those with the thinnest culm wall..."

Little Ginkgo felt fascinated and smiled happily.

森林的故事·竹子篇
The Story of the Forest · Bamboo Chapter

Morphological characteristics of *Pleioblastus distichus*

 Knowledge Point

Bamboo Plants: Bamboo plants are rich in species, consisting of more than 90 genera and over 1400 species. There are tree-like bamboos which can reach up to tens of meters in height and tens of centimeters in diameter, and herb-like bamboos which are only tens of centimeters high and a few millimeters thick. China has various bamboo species, with over 40 genera and over 600 species distributed around the country.

The Three Bamboo Families

China Bamboo Museum is the only bamboo museum in China. Visitors can not only see some rare bamboo varieties in the world here, but also learn about the thousand-year history of bamboo processing and utilization.

Little Ginkgo were immersed in the museum as there was so much for his eyes to feast on. Looking at the various bamboos, the term "three bamboo families" that Tony once mentioned flashed in his mind. Since it had puzzled him all along, he couldn't wait to ask Tony about it.

Tony did not answer immediately, but took Little Ginkgo to the bamboo root specimen showcase for observation.

"I know! This is bamboo rhizome, the underground part of a bamboo stem."

"That's right. If you look closely, you may find that these three specimens represent the three forms of bamboo underground stems: monopodium, sgmpodium amd amphipodiam. People usually identify species by the growth characteristics of plants. In terms of bamboo, such characteristics correspond to the three forms of underground stems, based on which bamboo can be divided into three categories: scattered, clustered and mixed bamboos."

"Then, which kind of bamboo does this belong to?" asked Little Ginkgo, pointing to a specimen.

"This is monopodial type. It grows into a scattered bamboo. Its bamboo rhizomes, long and thin, grow laterally in the soil. Buds will grow on the nodes of the rhizome, develop into shoots and come out of the earth, forming bamboo

Hear the story of bamboo

森林的故事・竹子篇
The Story of the Forest · Bamboo Chapter

groves in scattered state."

"Does this belong to a sympodial type?" Little Ginkgo pointed to a specimen and said, "I see clumps of bamboos, and they hug so tightly with each other that people can't get in at all!"

"That's right!" Tony said delightedly, "The sympodial underground stems form clustered bamboos. With extremely shortened bamboo rhizomes and very dense bamboo joints, the new bamboo shoots grown from terminal buds

stay close to the old culms. That's why the bamboo culms grow densely on the ground."

"Then, what about the amphipodial type?" Little Ginkgo couldn't wait to ask.

Tony guided Little Ginkgo to another specimen and explained, "The amphipodial underground stems have the characteristics of both monopodium and sympodium, so the part that grows above the ground includes not only the bamboo clumps with dense culms but also sparsely scattered culms, both of which form a bamboo grove. Such type of bamboo is thus called mixed bamboo."

"Now I understand what 'the three bamboo families' refer to! It turns out that various kinds of bamboos can be classified into three categories. It's much easier to identify!"

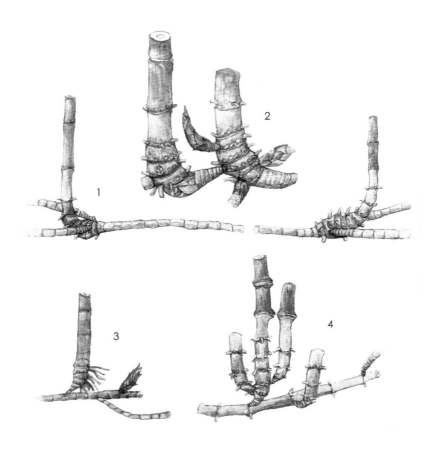

Rnizome types of bamboo
1. sympodial scattered subtype   2. sympodial cluster subtype
3. monopodial type   4. amphipodial type

 Knowledge Point

Bamboo Underground Stems: Bamboo underground stems can be classified into three categories: monopodium, sympodium and amphipodium, which develop into scattered, clustered and mixed bamboos respectively.

The Vine-like Climbing Bamboo

Little Ginkgo and Uncle Tony were visiting Xishuangbanna Tropical Botanical Garden. Little Ginkgo saw countless curtains hanging upside down on the tall arbors in the woods by the roadside, with fine green leaves. He said excitedly, "If these vines can be weaved together, we can make a natural swing."

Tony smiled and said, "These are not ordinary vines in the rainforest. They are bamboos."

"How could they be bamboos?" Little Ginkgo was very puzzled. In his impression, bamboos are tall and straight.

Tony led Little Ginkgo into the forest to observe carefully. They found that these bamboos were in clusters, with one to three centimeters in diameter. The culm was about 20 to 30 meters long, with dozens of short branches clustered at each node. Some clung to the arbors till their tops and hung out of the canopy, while others crouched themselves on the ground, stretching out.

Looking at the puzzled expression of Little Ginkgo, Tony explained, "This is a type of bamboo with unique appearance and of ornamental value. They are called 'the rattan bamboo'."

"They got this name just because they look like rattans?"

"Exactly. Can you find the biggest difference in shape between such bamboos and the vines you usually see?"

Little Ginkgo, who was ready for challenges, did not flinch. He observed the bamboo clump in front of him and tentatively answered, "They have smooth culms, but with no small tentacles."

Hear the story of bamboo

森林的故事・竹子篇
The Story of the Forest · Bamboo Chapter

"You are such a little science fan who observes carefully enough!" Tony gave him a thumbs up and said, "As you mentioned, the rattan-like bamboo does not have climbing organs like tendrils, so it grows up by clinging to the trunk. When reaching the top of the branch, it will hang down to the ground for lacking

Yes. Rattan-like bamboos do, but they don't have climbing organs like tendrils.

Are there any bamboos that grow by clinging to arbors?

external support. Therefore, people in some areas also call it 'the hanging bamboo'."

"There are about 5 genera and 20 species of Chinese rattan bamboos, including *Dinochloa, Melocalamus, Teinostachyum, Ampelocalamus, Neomicrocalamus*, etc. In addition, there are 15 semi-climbing bamboos or so, such as *Cephalostachyum, Leptocanna, Schizostachyum, Pseudostachyum, Bonia*, etc. Most of them are distributed in tropical areas or the low-altitude valleys with high temperature and frequent rain in southern tropics. Yunnan Province is the most concentrated area of rattan bamboos in China. Besides, Hainan Province and Tibet Autonomous Region, which are grouped as the Qiongdian Climbing Bamboo Area, serve as one of the five largest bamboo areas in China."

"One of the five? Then what are the other four bamboo areas?"

"I'll leave this question to you to answer by yourself." Tony patted Little Ginkgo on the head and assigned him an after-school homework.

 Knowledge Point

The Distribution of Bamboo: China is one of the central bamboo producing areas and the distribution of bamboo has obvious zonal and regional characteristics. According to the distribution of bamboo plants, there are five areas in the country: northern scattered bamboo area, Jiangnan alpine bamboo area, southwest mixed bamboo area, southern cluster bamboo area and Qiongdian climbing bamboo area.

Morphological characteristics of *Neomicrocalamus microphyllus*
1-2. a part of the culm, branches
3-4. ventral and dorsal surface of culm sheath

Bambusa ventricosa with Fat Waist and Round Belly

Recently Uncle Tony has brought back a bamboo bonsai which he was very fond of. Every time he took care of it, he couldn't help singing, "I would rather live without meat than live without bamboo. No meat makes people thin, and no bamboo makes people vulgar." Tony laughed, looking not merely like a Maitreya Buddha but also much like the bamboo in front of him.

After close observation, Little Ginkgo found that the bamboo in this potted plant had a peculiar shape. It had thin bamboo joints, but the internodes were short, thick and swollen, like the belly of Maitreya Buddha and the stacked Arhats. Little Ginkgo, amused by its funny and cute appearance, hurriedly pulled Tony and asked, "What kind of bamboo is it? Is it big belly bamboo?"

Tony was even more proud. He said in a Beijing accent, "That's right! It's the precious *Bambusa ventricosa*."

"I am wondering why it grows like this?" Little Ginkgo asked in a serious manner. His question interrupted Tony's self-indulgence.

"Usually, when the bamboo culm grows, the cell division and elongation activities on the same node basically go hand in hand, so the internodes are roughly in the same length. However, the cell division and elongation on the same node of *Bambusa ventricosa* grow at different times and with diverse speeds, which lead to deformed changes in the shape of the bamboo nodes. In some cases, the bamboo nodes are inclined to form a strange shape, thus growing into a 'monster' among bamboos which is used for garden landscaping and

Hear the story of bamboo

森林的故事・竹子篇
The Story of the Forest · Bamboo Chapter

enjoyment.

    Little Ginkgo scratched his head and came to a conclusion, "Isn't this an 'unintentional positive outcome'?"

    Tony applauded the witty summary of Little Ginkgo.

腰肥肚圆的佛肚竹
Bambusa ventricosa with Fat Waist and Round Belly

Bambusa ventricosa

1. shoots 2. branches and leaves 3. the outer surface of culm sheath
4. the inner surface of culm sheath 5. heteromorphic culm 6. normal culm

 Knowledge Point

Malformed Growth of Bamboo Joints: During the growth of bamboo, when the speed of cell division, elongation and enlargement discord with each other, the shape and arrangement of cells in the culm wall will be deformed, giving rise to the changed shape of bamboo culm, deformed internode overflow or swelling, and interactively tilted bamboo joints.

森林的故事·竹子篇
The Story of the Forest · Bamboo Chapter

Bamboo Inlaid with Gold and Jade

Nanjing Forestry University has a Baima Base Bamboo Plantation Garden, which is a base for bamboo research and teaching, academic exchanges and demonstrations, as well as tourism and leisure.

There are winding paths in the Bamboo Garden. Uncle Tony introduced the rich and diverse bamboo species in the garden as he walked. Little Ginkgo was listening and observing around.

Suddenly, a bamboo grove in the front left caught his eye. The bamboo rod was straight and the color was dazzling, like gold bars inlaid with pieces of jasper.

Before Little Ginkgo could ask a question, Tony introduced, "This is the *Phyllostachys aureosulcata* 'spectabilis'. As the name suggests, its culm is yellow, and a shallow green ditch is naturally formed at each branch and leaf which staggered with one another. You can go up there and verify."

Little Ginkgo, full of suspicion, got into the bamboo forest, observing one by one. Looking at the childish appearance of Little Ginkgo, Tony said: "The culms of young *Phyllostachys aureosulcata* 'spectabilis' are bright yellow, and then they gradually turn golden yellow, with green vertical stripes formed between nodes, the color yellow alternating with green."

"Why do they grow like this?" asked Little Ginkgo.

"This is related to the content of chlorophyll, carotenoid and anthocyanin in bamboo leaves and culms. They are like different pigments, the harmony of which causes changes in the color of leaves and bamboo culms. Similarly, there

Hear the story of bamboo

is the *Phyllostachys aureosulcata*, whose green culms are trimmed with golden stripes. Another typical example is *Phyllostachys nigra* (black bamboo) in the distance, which has green culms when it's young but one year later, purple spots will appear and finally the whole culm body becomes purple and black."

Little Ginkgo praised, "The changes are so magical in the slim bamboo culm. It seems that bamboo is also a brilliant colorist!"

森林的故事·竹子篇
The Story of the Forest · Bamboo Chapter

Phyllostachys aureosulcata **'Spectabilis'**

1. shoots　　2. branches and leaves　　3. the outer surface of culm sheath
4. the inner surface of culm sheath　　5. culm

 Knowledge Point

　　Pigment Changes in Bamboo Culms: The red, green and yellow colors of plant leaves, culms and fruits are often related to the type and content of chlorophyll, carotenoid and anthocyanin they contain. Studies show that, compared with all-green bamboo culms, changes happen to the genes that control pigments in some bamboos, which affects the deposition of chlorophyll in bamboo culms, resulting in different colors.

Bambusa multiplex

Little Ginkgo and Uncle Tony took a cruise to Sichuan Province. In the early morning, standing on the bow of the ship and leaning on the railing, they saw mist covering the fields on both sides of the Minjiang River. After the ship docked, they came to a holiday resort for a meal.

Walking into the courtyard, Little Ginkgo saw a clump of bamboo planted on the edge of the rockery. Their leaves were densely drooping and swaying, just like the posture of the plants from a distance. Little Ginkgo hurriedly sought advice from Tony.

Tony said affectionately, "This is the *Bambusa multiplex*, which is also called the Phoenix Bamboo."

"Its bamboo clumps look like the beautiful tail of a phoenix in shape, hence the name is Phoenix Bamboo. As for the name *Bambusa multiplex*, does it also possess special meaning?" Little Ginkgo went on.

"*Bambusa multiplex* is a clumping bamboo. One of the most notable features is that its bamboo rhizomes are closely united, and the new shoots grow around the mother bamboo, forming a cluster-by-cluster growth pattern. The bamboo rhizomes do not expand at random, and the bamboo shoots do not grow disorderly. Moreover, this growth pattern never changes."

Tony found that Little Ginkgo didn't seem to understand, and then explained, "The new shoot from this bamboo is just like a filial child, and the mother bamboo is like a kind mother who never dislikes the bamboo shoot in spite of the burden the child has brought. The mother is kind and the child is fil-

Hear the story of bamboo

ial. They grow together. Therefore, people have given it a sweet name — *Bambusa multiplex*."

Little Ginkgo suddenly remembered the knowledge point of the sympodial type, and drew an inference, "In this case, *Bambusa multiplex* should be sympodium! The name really matches it!"

"Exactly!" Tony repeatedly praised, "There are many interesting folk stories about *Bambusa multiplex*. You can find some to read in your spare time."

"I will always love and protect my mother, just like the new shoot." Little Ginkgo said silently in his heart.

母慈子孝的孝顺竹
Bambusa multiplex

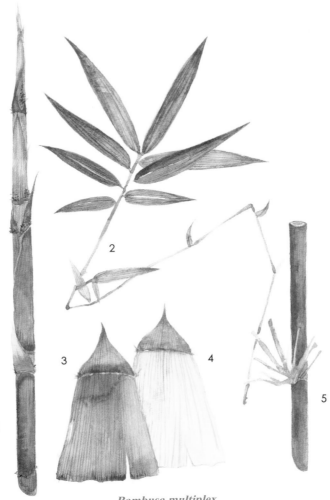

Bambusa multiplex

1. shoots   2. branches and leaves   3. the outer surface of culm sheath
4. the inner surface of culm sheath   5. culm and branches

 Knowledge Point

Clumping Bamboo: The underground stems of clumping bamboo are not slender bamboo rhizomes that go horizontally underground, but extremely shortened and densely knotted. The new bamboos grown from the terminal buds are close to the old culms, thus forming bamboo clumps with densely grown bamboo culms on the ground.

43

森林的故事・竹子篇
The Story of the Forest · Bamboo Chapter

Phyllostachys bambusoides 'Lacrimadeae' Sprinkled with Thousands of Tears

Uncle Tony treasures a bamboo flute, and often takes it out to play with. The surface of the polished bamboo flute is dotted with irregular spots, which is very unique.

Little Ginkgo couldn't help but ask, "Uncle Tony, is this flute very valuable? What kind of bamboo is it made of?"

Tony said, "Although the flute is an ordinary musical instrument, this flute is indeed quite precious, because the bamboo which it's made of is very precious. It is made of *Phyllostachys bambusoides* 'lacrimadeae' which is also called Bamboo of Xiang Concubines."

Little Ginkgo thought about what Tony said, "As for *Phyllostachys bambusoides* 'Lacrimadeae', we can understand it from its appearance; bamboo of Xiang Concubines? Perhaps there is a story, right?"

Tony introduced, "The *Phyllostachys bambusoides* 'Lacrimadeae' is distributed in all areas of the Yellow River and the Yangtze River basin. It is said that the *Phyllostachys bambusoides* 'Lacrimadeae' on Junshan Island in Yueyang has moire and purple spots, which are very much like tear stains. If the *Phyllostachys bambusoides* 'Lacrimadeae' is transplanted elsewhere, the stains will disappear in the second year. But if you move this *Phyllostachys bambusoides* 'Lacrimadeae' back to Junshan Island, it will be mottled again the next year."

"Ah, it's so amazing! Where did these spots come from?" Little Ginkgo asked curiously.

Hear the story of bamboo

"Bamboo experts have found that the spots on the bamboo are closely related to the soil and climatic conditions in which it grows. Their patterns are essentially the various plaques formed on the surface of the bamboo after the bacteria eroded its body. The new culm doesn't have spots initially, when it's nine months old, spots start to appear."

Little Ginkgo, who held his breath, breathed a sigh of relief and said, "Oh, so that's the case. Then why is it also called Bamboo of Xiang Concubines?"

Tony kept Little Ginkgo in suspense, "It is said that in the era of Yao and Shun...Why don't you go online and search for the myth by yourself?"

Little Ginkgo pretended to pout, but quickly turned on the computer and went to check the information! Tony smiled with relief.

森林的故事·竹子篇
The Story of the Forest · Bamboo Chapter

Phyllostachys bambusoides **'Lacrimadeae'**

1. culm   2. branches and leaves   3. the outer surface of culm sheath
4. the inner surface of culm sheath   5. shoots

 Knowledge Point

    Bamboo of Xiang Concubines: Bamboo of Xiang Concubines, also known as *Phyllostachys bambusoides* 'Lacrimadeae', is a famous ornamental bamboo since its culms are covered with brown moire and purple spots. Its culms can also be used to make handicrafts and bamboo wood.

Chapter 3 Bamboo Cultivation
No Bamboo is Vulgar

森林的故事·竹子篇
The Story of the Forest · Bamboo Chapter

Scientific Ways of Bamboo Breeding

The Chinese Arbor Day was coming. Little Ginkgo and the aunts and uncles of the Bamboo Research Institute went to the ecological garden to plant bamboos.

Sitting in the car, Little Ginkgo browsed the spring view along the road and asked, "Uncle Tony, how do you plant bamboos? Do you plant small bamboo seedlings in the way you plant trees?"

Tony smacked his lips and said with interest, "There are similarities and differences between planting bamboos and trees." Looking at the focused expression of Little Ginkgo, Tony continued, "There are several ways to plant bamboos. The first one is to grow seedlings, which means spreading bamboo seeds in the nursery and then transplant them after the seedlings are successfully raised; the second way is to bury the rhizomes to raise seedlings, which means burying the strong bamboo rhizomes that are dug up and plant them on the leveled land in the early spring or autumn."

"Then which way are we going to use today?" Little Ginkgo interrupted Tony eagerly.

When Tony was about to answer, the car had arrived at the destination. After getting out of the car, Little Ginkgo saw that there were many bamboos whose roots were tied into balls with straw ropes in the open space, and hurriedly pulled Tony forward to take a close look. These bamboos were lush and green, with well-proportioned internodes which were as thick as two fingers, but their tips had been cut off.

Little Ginkgo mumbled to himself, turned his head and asked, "Uncle

Hear the story of bamboo

Tony, where are the bamboo seedlings and bamboo rhizomes we are going to plant?"

"Today we will adopt the third method of planting." Tony said with a smile, "That is to select bamboos which not only grow robustly free from pests and diseases but also in moderate size as mother bamboos for planting."

No sooner said than done. Little Ginkgo and Uncle Tony started to get busy. Tony added as he were planting, "The division propagation method is suitable for scattered bamboo, while ramet propagation is applied to clustered bamboo."

Little Ginkgo worked very hard. Tired as he was, he straightened up and raised his hand to wipe the sweat from his forehead. Looking at the rows of newly planted bamboos in front of him, he seemed to see the picture of bamboo shoots sprouting next spring and forests flourishing in the following summer, and thus smiled heartily.

森林的故事·竹子篇
The Story of the Forest · Bamboo Chapter

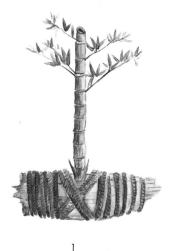

1　　　　　　　　　　　2

Transplantation ways of mother bamboo
1. transportation and binding of mother bamboo
2. erect support after planting mother bamboo

 Knowledge Point

Mother Bamboo Transplanting: Mother bamboo transplanting is one of the main methods of bamboo planting. It is divided into four steps: the first step is to select the mother bamboo, preferably the bamboo that is free from diseases and insect pests in its prime age. Usually, for *Phyllostachys edulis*, two-to-three-year-old bamboo is selected while for other types of bamboos, one-to-two-year-old bamboo is preferred; the second step is to dig out the mother bamboo, which needs to be dug up along with the bamboo culm and bamboo rhizome; the third step is to transport the mother bamboo. Before transportation, the bamboo top should be cut off, leaving only the bottom three to five trays of bamboo branches. This is not only convenient for transportation, but also reduces the water consumption of the mother bamboo, which is beneficial to the protection of the mother bamboo; the fourth step is to plant bamboo scientifically. When planting, keep the rhizome root stretching, and the depth of covering soil should be 3 to 5 centimeters deeper than the original submerged part of the mother bamboo. Meanwhile, a windproof bracket should be set up to prevent the bamboo culms from shaking.

Scientific Cloning of Small Bamboo Seedlings

At the Bamboo Research Center, Uncle Tony led Little Ginkgo to visit the bamboo tissue culture laboratory.

Putting on a white coat and walking into the clean laboratory, Little Ginkgo saw a row of metal shelves neatly lined up with small glass bottles, each with a small green bamboo seedling. Looking at these bottles of small green seedlings, he felt that they were very novel.

Tony introduced, "These small bamboo seedlings are all cultivated by cloning technology."

"What? Can bamboo also be cloned?" Little Ginkgo asked rhetorically.

Tony replied, "Cloning technology has been used in bamboo cultivation. People select excellent bamboos, use their buds, flowers or seeds, plant them in a medium with a prepared ratio after aseptic disinfection, and then put them into an environment with artificially controlled temperature, humidity and light, cultivate them. Eventually they will grow into complete small bamboo seedlings with buds and roots."

"It's so amazing! Most bamboos have a long flowering and fruiting cycle. Cloning bamboo seedlings is much more time-saving and faster than bamboo seedlings!"

Hear the story of bamboo

Tony said, "Exactly. By adopting this method, the production of bamboo seedlings will be free from the limitation of seasons. It also has the advantages of high yield and no occupation of arable land. Scientists can use new technologies to change bamboo genes and cultivate new species we need."

Little Ginkgo clapped his hands in amazement, "Technology promotes production. There is a new member of the bamboo family!"

Scientific Cloning of Small Bamboo Seedlings

Morphological characteristics of bamboo tissue culture

1. bud germinating   2. bud proliferation
3. bud growth   4. rooting

 Knowledge Point

Plant Cloning: Plant cloning is to cultivate plant tissues. Plant branches, leaves, buds and other parts of the organs or tissues contain all the information of the whole plant. Technology control can provide suitable temperature, light, air, heat, nutrition, hormones, etc.so that some organs or tissues such as branches, leaves and buds of plants can be cultivated to achieve rapid seedling growth. This is plant cloning.

Skills in Digging Bamboo Shoots

Hear the story of bamboo

It is the season for eating tender and delicious winter bamboo shoots. Upon hearing Uncle Tony's plan to dig winter bamboo shoots, Little Ginkgo jumped up with excitement. He quickly prepared the hoe, shovel, and machete with Tony, and set off merrily.

On the way to the bamboo grove Little Ginkgo asked, "Bamboo shoots are classified into spring shoots and winter shoots. Are they named because they grow in different seasons?"

"Yes. In addition to winter bamboo shoots, there are also summer bamboo shoots. But winter bamboo shoots taste the best, and they are known as 'golden

clothes and white jade, one of the best among vegetables'."

Upon hearing this, Little Ginkgo couldn't help wondering, "Will digging bamboo shoots reduce the number of bamboo shoots and new bamboos in the coming year?" He hurriedly sought confirmation from Tony.

Tony replied, "Don't worry, we have regulations for digging bamboo shoots. In terms of newly planted *Phyllostachys edulis*, it is forbidden to dig winter bamboo shoots within the first three years; after 3 years, when there are more than 180 bamboo plants per mu (about 1/6 acre or 667 square meters), bamboo shoots can be dug in a planned way. Reasonable digging can not only increase the economic income of forest farmers, meet the needs of consumers, but also ensure the normal growth of bamboos."

Little Ginkgo was relieved and hummed lightly.

Tony carefully observed around the bamboo and saw a crack on the sur-

face. He stepped on it lightly, then picked up the hoe to dig slowly. When the tips of the bamboo shoots were exposed, he used a shovel to remove the attached soil. Finally he used a machete to remove the winter bamboo shoots carefully.

Seeing that Tony was meticulous and methodical, Little Ginkgo asked, "It's so troublesome, why don't you dig up all the soil to find bamboo shoots?"

"What we used just now is called 'digging bamboo shoots by opening holes'. You are talking about digging bamboo shoots in a comprehensive way. Usually, it is combined with the loosening and fertilization of the bamboo forest in winter. When the bamboo forest is tended and reclaimed, the winter bamboo shoots are excavated, and it is advisable to turn the soil as deep as 20 centimeters or so. Experienced bamboo farmers also use the method of digging bamboo shoots along the rhizomes. But whichever method is used, the root rhizome of the bamboo must be protected from damage and breakage."

Little Ginkgo nodded earnestly and cooperated carefully. After a period of hard work, they harvested a basket of winter bamboo shoots and returned home with a full reward.

竹笋采控有技巧
Skills in Digging Bamboo Shoots

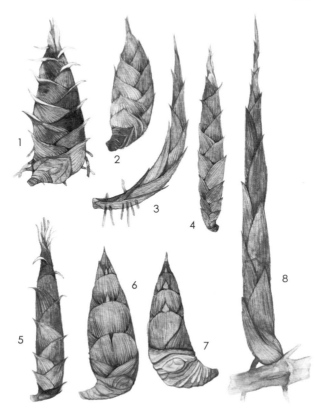

Morphological characteristics of several bamboo shoots

1. spring shoot of *Phyllostachys edulis*   2. winter shoot of *Phyllostachys edulis*
3. rhizome shoot of *Phyllostachys edulis*   4. shoot of *Phyllostachys sulphurea* 'Viridis'
5. shoot of *Phyllostachys glauca*   6. shoot of *Dendrocalamus latiflorus*
7. shoot of *Bambusa oldhamii*   8. shoot of *Chimonobambusa quadrangularis*

 ### Knowledge Point

Winter Bamboo Shoots Digging and Bamboo Forest Tending: Winter bamboo shoots usually refer to bamboo shoots of *Phyllostachys edulis* that have not yet sprung up from the soil. There are buds on each node of the bamboo rhizome and the physiological state of each bud is different. No more than 10% of the buds can develop into bamboo shoots under natural circumstances. Digging winter bamboo shoots can not only obtain delicious mountain delicacies, but it is also beneficial to promote the differentiation of other buds without affecting the density and uniformity of bamboo forests. Especially in winter, combining the excavation of winter bamboo shoots with bamboo forest reclamation, that is, digging up winter bamboo shoots and nurturing bamboo forests at the same time, can improve the economic and ecological benefits of bamboo forests.

森林的故事·竹子篇
The Story of the Forest · Bamboo Chapter

Phyllostachys edulis Tabbing and Filing

In midsummer, walking into the lush bamboo forest, Little Ginkgo suddenly felt a little cool. He looked up. The lush branches and leaves were like a giant green umbrella, blocking the scorching sun and casting coolness in the forest. He looked into the distance. The green bamboos, in the mottled light and shadow, were oily green, like a musical note beating in the woods.

Suddenly, Little Ginkgo noticed some red marks on the jade-like bamboo culm. When getting closer, he saw the number "8" marked with red paint. Little

Hear the story of bamboo

Ginkgo thought it was strange. He murmured, "Some people don't care about bamboos, so they use knives to engrave their names on them. But who wrote the red lacquer number?"

He carefully looked around the bamboos and found that many bamboo culms had different numbers written on them, besides "8", there were also "9", "7" and so on. He asked in confusion, "Uncle Tony, what do the numbers on the bamboos stand for? Are they mysterious symbols? And who wrote them?"

Tony replied, "These are the marks made by the workers. For such a large-scale forest farm, in order to strictly control the age of bamboos when logged, the forest workers will mark the bamboo culm with the year of its shoots during July and August. If it is the first year to mark, each bamboo in the bamboo forest is usually numbered with the year, which is called massive marking. With these numbers, each bamboo has its own personal file. In this way, the workers are informed when felling the bamboo."

"7, 8, 9" Little Ginkgo murmured, and his mind flashed, "I see! This means the year, right? The number "8" means that the bamboo shoot broke through the soil in 2018. Right?"

Tony applauded, "How smart you are! With this label, workers can cut down according to the number, which can ensure a reasonable age structure of bamboo forests. By adhering to the principle of 'cutting the old and keeping the young, cutting the dense and leaving the sparse, cutting the weak and leaving the strong', bamboo forests can grow healthily and can be used all the time!"

Little Ginkgo smiled and said, "It's not Little Ginkgo who is smart, but the experts in bamboo management!"

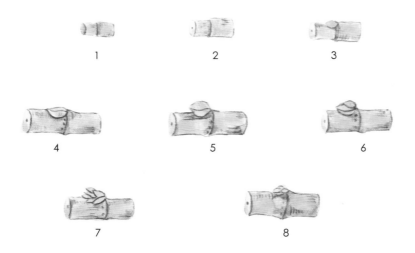

Changes of the rhizome age and lateral buds of *Phyllostachys edulis*
1-2. the bud at the end of a rhizome
3-4. The bamboo rhizome node at young age (1-2 years old)
5-6. The bamboo rhizome node at robust age (3 to 6 years old)
7-8. old bamboo rhizome(more than 7 years old)

 Knowledge Point

The Life Span of a Bamboo: The age of bamboo culm is called bamboo age. A bamboo is unearthed from a bamboo shoot to form a new-born bamboo, from a young age, a strong age to an old age till it dies. The life span of a bamboo varies from a couple of years to more than ten years.

Cutting Bamboo Tips to Resist Snow Disaster

In the golden autumn of October, Little Ginkgo and Uncle Tony went to Shitang Bamboo Sea for the weekend. Sitting in the teahouse for a break, Little Ginkgo found that the bamboo forest on the hillside beyond the lake was swaying greatly. Taking a closer look, he found that several people were hooking bamboo poles and cutting off the bamboo tips.

"Some people are destroying the bamboo forest!" Little Ginkgo shouted anxiously, pointing his finger forward.

Tony looked in the direction of Little Ginkgo's finger, tilted his head and tried to comfort him, "Don't worry! They are not destroying, but managing the

Hear the story of bamboo

bamboos by cutting off the tips."

"Why did they do that?" asked Little Ginkgo.

"Since the bamboo leaves are lush, in case of a heavy snow in winter, a large amount of frozen snow and icicle will accumulate on the branches and leaves, making the bamboo top-heavy, and then disasters such as turning over, breaking, lodging, etc. will occur, leaving a great impact on the growth and yield of bamboo forests. Cutting off the tips can prevent and reduce the hidden danger."

Little Ginkgo was still worried, and asked Tony to lead him to the scene to find out what was going on. He saw the forest ranger holding a sharp hook knife, aiming at the top of new bamboos, and quickly shaving off the tips and lateral branches.

Tony went on, "Cutting off the tips of *Phyllostachys edulis* prevents disasters and increases production. Removing the tips of bamboos will greatly enhance their ability to resist wind, frost and snow. After hooking, the bamboo forest has better light transmittance and stronger photosynthesis, which is beneficial to promote the growth of bamboo leaves, rhizomes and shoots, so that increased yield and more straight bamboo culms will be achieved."

After hearing Tony's explanation, Little Ginkgo smiled with relief. In the meantime, he also blushed for his misunderstanding of others.

 Knowledge Point

Tip Hooking: Tip hooking means using a sharp knife to hook off the tip of new bamboos. Its main function is to reduce the damage that snow and icicle bring to the bamboo forest in winter and early spring. Tip hooking can fall into three types according to the operating season: first, moldy tips, which are usually cut off during plum rain season in June; second, tips of dog days, which are hooked in hot summer from July to August; third, tips of White Dew, which are hooked around White Dew in early September.

Chapter 4 Utilization

No Residence, No Bamboo

森林的故事·竹子篇
The Story of the Forest · Bamboo Chapter

One Pole as a Boat

During the summer vacation, Little Ginkgo and Uncle Tony traveled to Guizhou Province.

"Wow! Uncle Tony, look!" Little Ginkgo suddenly shouted in surprise. Looking in the direction of Little Ginkgo's finger, Tony saw a man stepping on a thick bamboo, holding a bamboo pole as a paddle and slowly sliding on the water. "Oh, this is the famous unique skill of single bamboo drifting in the Chishui River Basin. The Red Army used to cross the Chishui River four times. During

Hear the story of bamboo

64

that period, they used the single bamboo as a boat to cross the river." Tony said with a smile.

"How can a bamboo support a person to float on the water?" asked Little Ginkgo curiously. Tony did not answer immediately, but smiled and touched Little Ginkgo's head.

Tony led Little Ginkgo to the river bank. On the beach, there was a big bamboo. "This is the bamboo used for the bamboo drifting we just saw." said Tony.

Little Ginkgo squatted down and looked at it carefully. It seemed that the big head of this bamboo was as thick as his own thigh, and it was about two stories high. Other than that, there seemed to be nothing special about it. Can this

The Story of the Forest · Bamboo Chapter

carry the weight of a person?

Tony also crouched down and reminded Little Ginkgo, "Will you count how many nodes there are on this bamboo?" Little Ginkgo took action immediately.

Tony went on saying, "Each partition between two nodes is an airtight chamber, like a small floating ball, which generates enough buoyancy to lift a person." Then in the same breath he asked, "Do you think any bamboo can be used for bamboo drifting?"

"No way. I think it must be thick and long enough, like this bamboo." Little Ginkgo observed it carefully for a while before giving a tentative answer.

Tony nodded in satisfaction and said, "That's right! If the bamboo is thin and short, the buoyancy will naturally decrease, and it won't be able to support a person."

Tony added, "Since this place is prolific of big bamboos, bamboo drifting has become a local folk stunt. In the 2011 ethnic minority traditional sports meeting, bamboo drifting was listed for the first time as a sport."

"It's amazing!" Little Ginkgo's eyes shone with admiration.

 Knowledge Point

Bamboo Culm Structure: Bamboo culm is the main body of a bamboo, including culm neck, culm base and culm. The culm is composed of culm rings, sheath rings, intranodes, septa and internodes, among which the first three are called nodes. The partition between two nodes is called internode. The internode is usually hollow, and there are bamboo partitions between the nodes. There are significant differences in the size, number of nodes, and length of internodes among different bamboo species.

一竿为舟的竹
One Pole as a Boat

Morphological characteristics of *Phyllostachys edulis*
1. shoot 2. branches and leaves 3. the outer surface of culm sheath
4. the inner surface of culm sheath 5. culm

森林的故事・竹子篇
The Story of the Forest · Bamboo Chapter

Bamboo for Landscaping

Hear the story of bamboo

Having followed Uncle Tony to many places and visited many gardens, Little Ginkgo found that there were bamboos everywhere. He asked Tony, "Why do people like to plant bamboos in their gardens?"

Perhaps the question was too difficult to answer, Tony thought for a while before reply, "China is 'the Kingdom of Bamboos'. There are a wide variety of

Why do people to plant bamb in their garde

bamboos, which are widely distributed and easy for people to use."

Inspired by Tony's words, Little Ginkgo cut in eagerly, "The green bamboos are in a variety of shapes. They are a combination of beauty in form, color, and artistic conception, and are of high ornamental value."

Tony gave a thumbs up and said in the same literary tone as Little Ginkgo, "Yes! Different bamboo species have different shapes. Some are less than a foot in height, and some are towering in the sky; some stand alone, and some gather into clumps; some leaves are as thin as needles, and some leaves are as big as palms..."

 森林的故事·竹子篇
The Story of the Forest · Bamboo Chapter

Tony's words became more literary, "Building a garden with bamboos, regardless of the scattered bamboo shadows formed by itself, or using bamboos to create, borrow, and block a scene, or using bamboos to decorate, frame, and move a scene, can form a picturesque scenery with a variety of styles."

"Let me say something, let me say something!" Little Ginkgo was also in high spirits and opened the conversation, "*Bambusa ventricosa* and *Phyllostachys aureosulcata* can be planted for viewing poles, colors, and shapes in the courtyard; *Phyllostachys edulis* can be planted to separate spaces; evergreen *Bambusa multiplex*, *Phyllostachys glauca*, *Phyllostachys nigra*, *Phyllostachys edulis* 'Kikko-chiku' can be chosen to decorate corners; as for the bamboo fence, clumping bamboo or mixed bamboo is the best choice."

Tony was so excited that he said repeatedly, "It's worthwhile to have taken you to so many places!"

 Knowledge Point

Chinese Garden and Bamboo: Chinese people love bamboo for its evergreen color, fearlessness of severe cold, and modesty. They take bamboo as the standard of being a person. Horticulturists select to plant and appreciate the bamboo according to their forms of culms, branches, leaves and shoots, thus creating the characteristics of Chinese bamboo gardens.

Bamboo that Is as Strong as Steel and Iron

It is a beautiful village of the Dai nationality. The unique bamboo buildings of Dai people, hidden in the dense *Bambusa multiplex* and oil palm forests, look so peaceful and beautiful. The tour guide introduced that the Dai area is rich in bamboos, and the villagers like to use local materials. They use thick bamboos as keel to build the frame, use bamboo strips as walls, use bamboo strips or wooden boards as floors, and then lay grass on the roof. And then, a simple and ventilated bamboo building is completed.

Having left the bamboo building, Little Ginkgo asked Tony, "Why can bamboos swaying with the wind be used to create a two-story building where people, storage, and livestock can be housed safely and simultaneously?"

"This is all because bamboo has a special body that is 'as strong as steel and iron'." Tony chatted with Little Ginkgo as they walked.

"Bamboo's special body stems from the way it grows. Bamboo grows from the outside to the inside, so the surface is the hardest. Bamboo generally grows very tall, but the bottom of the culm is only slightly thicker than the top. Only if they can bear their own weight and have high tensile strength can they sway with the wind without bending. When you break the bamboo culm, you will find that there are many filaments that are hard to pull apart, that is, the tough bamboo fibers, which tightly connect the entire bamboo culm, just like the reinforced concrete used to build high-rise buildings."

Little Ginkgo asked, "What is the impact of the hollow and multi-knot

Hear the story of bamboo

structure of bamboo?"

"Bamboo's hollow structure makes its bending resistance much stronger than that of the solid structure with the same cross-sectional area. Bamboo's knots are like hard 'beams', which play a supporting role and reinforce layer by layer." Tony said with a smile, "Now, you see. Bamboo is a building material with high strength and good stability!"

Little Ginkgo couldn't help envying bamboo after listening to Tony's words. He praised in his heart, "Bamboo is really amazing!"

 Knowledge Point

Mechanical Characteristics of Bamboo: Bamboo has excellent mechanical properties such as strong stiffness and high strength, and is a good engineering structural material. Its static bending strength, tensile strength, elastic modulus and hardness are about two times higher than those of ordinary wood.

Domestic Use

Having nothing to do, Uncle Tony and Little Ginkgo sat around the tea table, drinking tea and talking about the role of bamboo in human life.

Little Ginkgo kept talking, "We cannot live without bamboo, such as the bamboo mat, bamboo bed for sleeping in summer, bamboo chopsticks, toothpicks, bamboo rocking chairs we sit on, the bamboo basket my grandma uses to buy vegetables, fishing rods of my father, bamboo brooms used by the sanitation workers, the bamboo bookshelf in my room, and…"

Tony nodded, knocked the tea table with his hand, and said mysteriously, "Look! What material is this table made from?"

Little Ginkgo widened his eyes. He touched, looked around, knocked and lifted, but he still couldn't figure out the material. He said mischievously, "You haven't taught me the knowledge of wood, how would I know? It's a kind of wood anyway."

"Ha, it is also made from bamboo!"

Little Ginkgo really couldn't figure out that bamboo had anything to do with the smooth, beautiful and natural color tea table in front of him. He said suspiciously, "Can't it be true? It's completely different from the bamboo bookshelf in my room. I can't see any trace of bamboo at all."

"It is modern bamboo furniture. Most of the traditional bamboo furniture is directly made of bamboo by hand. However, their structures are not solid enough, and because of the rough process, the products are prone to insects, mold, and will crack and deform if affected by certain environment."

Hear the story of bamboo

Hearing what Tony said and thinking about the appearance of the bamboo bookshelf in his room, Little Ginkgo nodded in approval.

"In order to overcome these shortcomings and give full play to the role of bamboo, people now turn bamboo into boards and then make furniture. The tea table is made of bamboo board."

Tony got up and led Little Ginkgo to appreciate the other pieces of bamboo furniture that had just been added to the house, and said, "The bamboo board can be used at will, realizing the perfect combination of furniture fashion and environmental protection, elegance and comfort. It has many advantages: no dust, no dew condensation, easy to clean, and borer-proof."

"It is the perfect combination of nature and technology!" Little Ginkgo tutted in admiration.

 Knowledge Point

Bamboo Wood-based Panel: Bamboo wood-based panel, which is mainly made from bamboo, is a kind of plate and square material, through a series of physical and chemical treatment and mechanical processing, and glued and pressed under a certain temperature and pressure. It eliminates the weaknesses of bamboo such as anisotropy, discontinuity and being easy to crack, and has excellent physical properties.

Bamboo that Transforms into Soft Silk

After visiting the Bamboo Museum, Uncle Tony and Little Ginkgo walked into the souvenir shop at the exit. The store displayed a dazzling array of bamboo products. In addition to bamboo fans, bamboo bowls, bamboo cups, and musical instruments made of bamboo, there were also towels, socks, clothing and bedding. These items immediately attracted Little Ginkgo.

"Bamboo clothes?" Little Ginkgo blinked at Tony, "Can bamboo really be made into clothes?"

Tony smiled and said, "Yeah. Bamboo fibers extracted from bamboo, like cotton, flax, wool, and silk, can be spun and woven to make clothes."

Little Ginkgo was stunned, and reached out to touch the bamboo fiber towel, bamboo fiber socks, and bamboo fiber clothing. He felt soft, smooth and cool. But he still felt confused and asked, "Usually everyone likes to wear cotton and linen clothes, so what merits do bamboo clothes have to attract people?"

"This has to be answered from the microstructure of bamboo fibers. Observed under a scanning electron microscope, it can be seen that the bamboo fiber has slender holes and surface grooves. This porous structure allows it to have excellent hygroscopicity and dehumidification, so as to automatically adjust the humidity of the human body and achieve the effect of being warm in winter and cool in summer. There is also a unique substance in bamboo, which has natural antibacterial, anti-mite, anti-odor and anti-insect functions. These are all its ad-

Hear the story of bamboo

vantages over cotton and flax fiber products."

Tony took Little Ginkgo's hand and said, "Let's go! Uncle Tony will take you to choose a dress, so you can have a closer contact with bamboo!"

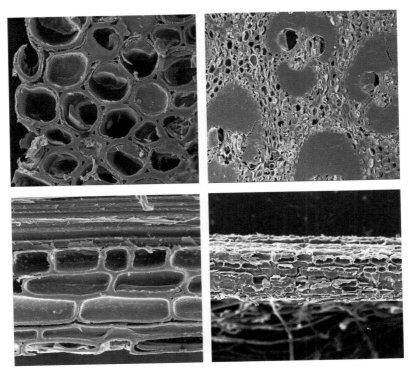

Electron microscope micrograph of bamboo fiber

Knowledge Point

Bamboo Fiber: Bamboo fiber refers to cellulose fibers extracted from naturally grown bamboo. The structure of the slender hollows and surface grooves of bamboo fiber enhances the capillary effect, so that it has excellent hygroscopicity and dehumidification, and becomes the first choice for textile processing. According to different processing methods, bamboo fiber for textile is divided into three categories: bamboo raw fiber, bamboo pulp fiber and bamboo carbon fiber.

森林的故事・竹子篇
The Story of the Forest・Bamboo Chapter

Bamboo that Can Be Used to Make Good Medicine

Hear the story of bamboo

In early spring, Uncle Tony brought back a few green bamboos from the bamboo forest and set up an oven in the yard. Seeing this, Little Ginkgo was delighted and asked behind Tony, "Uncle Tony, are we going to make bamboo rice?"

Tony looked at Little Ginkgo's greedy look, and said with a smile, "We are not going to cook food today, but to make a good medicine—fresh bamboo juice."

"Make medicine?" Little Ginkgo was very curious.

Tony sawed the bamboo pole into a bamboo tube of about 50cm, slightly longer than the length of the oven, split it in half and removed the bamboo knots in the middle. Then he placed the bamboo slices horizontally on the oven, stirred

slightly the charcoal fire, and placed several glasses directly under both sides of the bamboo slices.

After ten minutes of grilling, the bamboo slices began to sizzle, small bubbles appeared at both ends, and liquid gradually leaked out, just dripping into the glass below.

Little Ginkgo stared at every step of Tony's operations with full attention. Tony said slowly while grilling bamboo slices, "The juice we are collecting in the glass is the bamboo juice, which has the effect of clearing away the lung-heat, resolving phlegm and facilitating excretion. It is a good prescription recorded in *Chinese Materia Medica*. It is also a backup medicine that ordinary people make at home."

Tony took a cup, poured in some bamboo juice, mixed it with about three times of the mineral water, and handed it to Little Ginkgo. Little Ginkgo tried to take a sip. The bamboo juice was a sweet taste. Instantly, he felt a little coolness

森林的故事·竹子篇
The Story of the Forest · Bamboo Chapter

sliding down his throat.

Seeing Little Ginkgo enjoying himself, Tony continued, "From root to tip, many parts of bamboo have medicinal value. For example, *Phyllostachys glauca* leaves can remove heat and dryness, refresh and enrich the saliva and diuresis; bamboo shavings is cool in nature, and it clears heat and resolves phlegm, removes vexation and stops vomiting; young bamboo leaf, that is, the young leaves of tender bamboo leaves that are rolled but not opened, has the effects of clearing heart, removing annoyance, relieving summer heat and thirst..."

"Uncle Tony, slow down, please!" Little Ginkgo rubbed his head, "I can't catch you!"

Tony smiled and continued as if he was talking a cross talk, "And bambusicola, a lump of solidified secretions formed after the bamboo is injured, has the effect of clearing away heat and detoxifying, and calming the mind; bamboo culm membrane, the coating film inside the *Phyllostachys sulphurea* culm, can cure throat hoarseness and cough; bamboo essence, the liquid juice in the lumen of young bamboos, can cure sweat spots..."

Little Ginkgo held the cup and listened in fascination. At that moment, he had already extremely worshiped bamboo. He said to Tony, "No wonder you often say 'the whole body of bamboo is a treasure, and it can be used as medicine to cure diseases'."

While nodding approvingly, Tony poured the bamboo juice from the glass into a glass jar and sealed it for storage in the refrigerator.

 Knowledge Point

Bamboo Juice: Bamboo juice is the liquid from the stems of gramineous plants such as *Phyllostachys edulis* after being grilled by fire. It has the effects of clearing away the lung—heat, resolving phlegm and facilitating excretion.

Air Purification

Little Ginkgo found an interesting bag in Uncle Tony's car. The beige cloth bag looked like a sachet. It rustled when he touched it. It seemed that there were many small pearl particles in it.

"Uncle Tony, what is this?" asked Little Ginkgo curiously.

Tony rubbed the small bag and said, "This is a bamboo charcoal bag, filled with bamboo charcoal, which is used to absorb odors in the car and purify the air."

Little Ginkgo picked it up and smelled it, a faint scent of bamboo swept across the tip of his nose, and the air he breathed seemed to be a lot fresher. "My God, it's almost as amazing as planting a bamboo in the car! How did this happen?"

Seeing Little Ginkgo's cute face while he breathed greedily, Tony said with a smile, "People use a special process to turn the green bamboo into black and shiny bamboo charcoal. There are countless cells invisible to the naked eye in the vascular tissue of bamboo. The water in the cells is extracted during the charcoal burning process, leaving the carbonized cell wall, forming a porous structure, making bamboo charcoal the best medium for purifying the environment. Compared with charcoal, bamboo charcoal has higher porosity and its surface area is two to five times that of charcoal, so its adsorption is much more powerful than that of charcoal."

Holding the small but powerful bamboo charcoal bag, Little Ginkgo asked, "I know that activated charcoal made of wood can purify waste water, so can bamboo charcoal do it?"

Hear the story of bamboo

"Of course!" Tony said proudly, "Bamboo charcoal is very useful. It can be used for drinking water purification, soil improvement, wastewater treatment, residential humidity regulation, deodorization and so on. It can also be used to make bamboo charcoal fuel and bamboo charcoal fiber. Isn't it awesome?"

Little Ginkgo widened his eyes, "It's so amazing! Not only bamboo forests can purify the air, but bamboo charcoal is also an expert in purification."

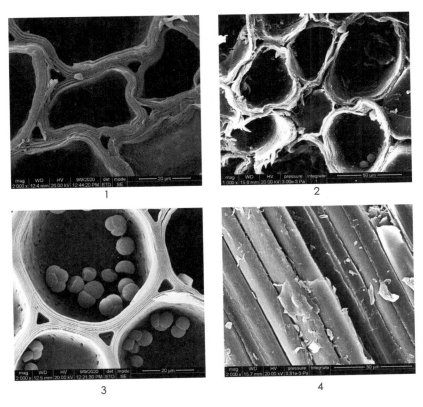

Electron microscope micrograph of bamboo charcoal
1-3. cross section   4. longitudinal section

 Knowledge Point

Bamboo Charcoal: Bamboo charcoal is the main product of bamboo pyrolysis. Mature bamboos of more than four years old are usually selected and burned at high temperature without oxygen to make bamboo charcoal. There are mainly two ways of making bamboo charcoal: dry distillation pyrolysis and direct firing in the earth kiln.

森林的故事・竹子篇
The Story of the Forest · Bamboo Chapter

Bamboo that Transforms Air into Melody

Walking out of the concert hall of the Chinese Orchestra, Little Ginkgo was still immersed in the melodious music. He couldn't help but describe his listening experience to Uncle Tony: the clear sound of the flute, the low sound of Xiao (a vertical bamboo flute), the soft sound of Huqin (a two-stringed Chinese violin) ... All melodies were so beautiful.

"Do you know that most of these ethnic musical instruments are made of bamboo." Tony said, "Bamboo is one of the eight instruments. *Records of the Historian* in ancient times, a large number of bamboos grew in the Yellow River basin, so people at that time began to choose bamboo as the material to make

Hear the story of bamboo

wind instruments."

"Why do people think of using bamboo to make musical instruments?"

"Probably because bamboo has a round and hollow structure with uniform pipe wall and moderate density, which makes it have good vibration and crisp sound. In addition, the universality of growth makes it easy to obtain, and it is easy to process."

"When I go home, I will find a bamboo pole, and make a bamboo flute to play like those people in the TV series." Little Ginkgo said with a laugh.

"It is not a simple thing to make musical instruments from bamboo. To make an obscure bamboo growing in the bamboo forest into a musical instrument that makes melodious music, it all depends on a pair of skillful hands of the craftsman. It requires selecting the bamboo, drying in the shade, peeling, broiling the bamboo, scalding the holes, tuning, and polishing ... In total, there are more than 80 procedures."

森林的故事・竹子篇
The Story of the Forest · Bamboo Chapter

Little Ginkgo stuck out his tongue, regretting what he had rashly said.

"Take making a flute for example. The material should be selected from a bamboo of more than three years old, and the pipe wall of bamboo should neither be too thin nor too thick. If it is too thin, the sound will explode and float; if it is too thick, the sound will be dull, so the proper wall thickness is generally a little more than two millimeters. Also, not all parts but only one section of a bamboo can be used as wind instruments."

"In addition, the location of bamboo flute holes is also extremely important. After all, bamboo grows naturally and is not perfectly standard cylindrical, so holes are usually opened on the facade of the bamboo, that is, the most prominent side, which is the most convenient location to play. What's more, the most significant thing is tuning. Tuning is credited to the craftsman's sensitive hearing. When fine-tuning the flute holes, if the craftsman polishes a little more carelessly, the intonation of the flute may be wrong, and the hard work of many days will be wasted." said Tony.

Unconsciously, Tony led Little Ginkgo to the backstage. Here, Little Ginkgo saw various musical instruments such as flute, Xiao, Sheng (a reed pipe wind instrument), and Yu (an ancient 36-reed wind instrument), all of which were made of bamboo!

 Knowledge Point

Bamboo and Musical Instruments: Bamboo has an indissoluble bond with Chinese culture and art, and its role in music and art is particularly prominent, which is closely related to the particularity of bamboo structure and the extensive distribution. Chinese musical instruments made of bamboo are too numerous to mention, such as flute, Xiao, Sheng, Zheng (Chinese zither), Yu, and so on.

Chapter 5 Culture
Poetic Bamboo

森林的故事·竹子篇
The Story of the Forest · Bamboo Chapter

Bamboo in Traditional Chinese Painting

Hear the story of bamboo

Little Ginkgo visited the Palace Museum with Uncle Tony in Beijing. He was deeply attracted by the lifelike traditional Chinese paintings.

In the painting in front of them, the bamboo joints were tough and vigorous, the bamboo leaves were thin in shape. The background was the rugged rocks outlined with light ink. Accompanied by huge rocks was the vigorous bamboo swaying in the breeze, with an air of pride. Little Ginkgo felt that the bamboo was so vivid that he could not help chanting loudly, "It won't let go its

bite on the green hill, as it is rooted in the boulder crack."

Tony hurriedly pulled Little Ginkgo's sleeve, telling him to lower his voice, and then said, "This is Zheng Banqiao's painting *Bamboo and Rocks in Ink*. Although the composition is simple, the brushwork is very stable, showing the nobility, simplicity, elegance and tenacity of bamboo."

Little Ginkgo asked, "Bamboo is a very common plant in nature, but all the way down here, there are so many paintings of bamboos. Why do many people like to draw bamboos?"

Tony said, "In Chinese traditional culture, people like to express emotions through describing concrete objects. Plum, orchid, bamboo and chrysanthemum are deemed as 'four gentlemen', which are common themes for poetry and paint-

ing."

"Bamboo has a long history as the subject of Chinese painting. According to written records, some people drew bamboos in the Three Kingdoms Period and Jin Dynasties, and then the technique of bamboo drawing developed in the Southern and Northern Dynasties, Sui, Tang and Five Dynasties. In the Tang Dynasty, the method of directly painting bamboos in ink appeared. By the Yuan Dynasty, there were as many as 50 or 60 painters famous for their bamboo paintings. After the Ming and Qing Dynasties, the techniques and forms of bamboo painting became more diverse, so that bamboo painting has been greatly developed."

While talking and walking, Tony and Little Ginkgo stopped in front of a bamboo painting drawn by Su Shi. Tony pointed to it and asked heuristically, "When drawing a bamboo, you must have an image of it in your heart, so that you can do it in one take. This is called what?"

"Having a complete bamboo on one's mind or having a ready plan in mind!" Little Ginkgo blurted out.

 Knowledge Point

Bamboo Culture: China is the country with the richest bamboo resources in the world. Bamboo is one of the important forest resources in China, whose species, coverage and output account for about one third of the world's total. Therefore, China is known as "the Kingdom of Bamboo". As a country with a long history, China's bamboo culture is also rich and splendid. It is an important part of Chinese national culture, and it is also the essence and treasure of Chinese culture.

Bamboos in Ancient Poems

Chinese Bamboo Town Poetry Contest will be broadcast live on TV tonight. Before the show started, Little Ginkgo had already sat in front of the TV with Uncle Tony.

Accompanied by a piece of passionate music, the host announced, "The poetry game Feihualing" of depicting bamboos now starts!" Little Ginkgo put his heart in his throat, nervously watched the contestants. "Outside the bamboo grove appeared a few branches of peach blossom. When spring has warmed the stream, ducks are the first to know." The challenger began the first round.

"Amid the boom of firecrackers, a year has come to an end. And the spring wind has wafted warm breath to the wine." The defending champion said calmly.

"Washerwomen's laughter can be heard in the bamboo grove on their way back. The lotus stirs under the fishing boat." "When the wind blows across the river, it makes waves of a thousand feet. When the wind blows into the bamboo forest, ten thousand bamboos are tilted." "When the snowflakes fluttered into the window, I was sitting in front of the window watching the green bamboo branches covered with snow." "When seeing pines and bamboos, I feel as if I were in the mountain." and the like.

During the advertisement time, Little Ginkgo could not help holding Tony's hand and said in a low voice, "Uncle Tony, there are so many ancient poems about bamboos! It seems that poets like bamboos very much!"

Tony responded, "Yes, the ancient literati loved bamboos. The bamboo has culms, branches, joints, and leaves, which is hollow inside and straight outside,

Hear the story of bamboo

森林的故事·竹子篇
The Story of the Forest · Bamboo Chapter

and not afraid of the cold. Therefore, it has always been the object for expressing aspirations in poetry and painting, and bamboo culture has also become an important part of Chinese national culture. According to statistics, there are more than 14000 ancient poems, lyrics, poems and songs about bamboos."

Little Ginkgo was surprised. At this time, the fierce poetry game went on again.

"On bamboo horseback came the lad, circling around the well hedge, and plucking up green plums for his maid."

"My bamboo cane and straw sandals are better than riding. The tempest is anything but forbidding. I in a coir raincoat can spend my life striding."

"One can eat without meat, but one can't live without bamboo."

"There are thousands of green bamboos in front of the window, but people who don't know me can't see tears on them."

"..."

After several rounds, Little Ginkgo was filled with emotions from bamboo-praising poems throughout history.

 Knowledge Point

Natural Distribution of Bamboo in China: Bamboo is widely distributed in China, from Hainan Island in the south to the Yellow River basin in the north, and from Taiwan Province in the east and to the lower reaches of the Yalu Tsangpo River in Tibet in the west, covering from 18 degrees to 35 degrees north latitude and 92 degrees to 122 degrees east longitude. Among them, the south of the Yangtze River is the largest area for most bamboo species to flourish.

森林的故事・竹子篇
The Story of the Forest · Bamboo Chapter

Bamboo Slips for Writing

The museum held a cultural experience activity to introduce bamboo slips. Little Ginkgo was very active, coming with Uncle Tony to wait early in the morning.

The guide slowly opened a roll of bamboo slips on the booth and said, "Bamboo slips are bamboo slices used for writing in ancient times. Most slips are made of bamboo slices, and on each slice is a column of characters. After an article is written, all the bamboo slices are compiled and linked, which is known as 'Jiandu' (bamboo slips). It was an important writing material from the Warring States Period to the Wei and Jin Dynasties."

Little Ginkgo whispered, "Is it true that after Cai Lun improved paper-making in the Eastern Han Dynasty, people no longer used bamboo slips?" Tony smiled but did not answer and asked Little Ginkgo to pay attention to what the guide said.

"Since the appearance of paper, bamboo slips had still been used together

Hear the story of bamboo

with paper for hundreds of years. Writing on bamboo slips had not been out of use until the end of the Eastern Jin Dynasty. Later, in the Tang Dynasty, some history books used the phrase 'too numerous to inscribe on all bamboo strips' to describe a person's heinous crime, which shows that the use of bamboo slips has a long history."

As soon as the introduction was over, Little Ginkgo asked Tony, "Why did the ancients choose bamboo as writing material?"

Tony explained, "Bamboo is divided into the interior yellow skin, the bamboo flesh or meat, and the exterior green skin from the inside out. The interior surface is easy to get ink on and the ink is not easy to be erased. In addition, it can grow into forests in one year, so it is easy to obtain materials. Bamboo slices have compact texture, minute color difference, and strong toughness, so they are easy to be processed. These advantages prompted the ancestors to choose bamboo as writing material."

The final experience activity is to let everyone make bamboo slips. First, cut a symmetrical green bamboo culm, and cut the bamboo culm into evenly wide bamboo slips. Second, broil and dry, and then polish and thread those slips. After those procedures, a string of bamboo slips is considered finished. Finally, favorite characters can be written on it.

Easier said than done. Little Ginkgo did it in a flurry, which made him feel the difficulty of making bamboo slips by the ancients. Furthermore, he also understood the great contribution of bamboo in carrying Chinese civilization.

 Knowledge Point

Bamboo Wall: The cylindrical shell of the bamboo culm is called the bamboo wall, which is the general name of the epidermis to the pulp cavity of the bamboo culm. The bamboo wall is divided into three layers from outside to inside: the exterior surface of green skin, the bamboo flesh or meat, and the interior surface of yellow skin.

森林的故事・竹子篇
The Story of the Forest · Bamboo Chapter

History of Firecracker

    Chinese New Year's Eve in the countryside is really boisterous! Little Ginkgo and his friends set off firework and firecrackers in the yard.

    "One year has passed in the sounds of delightful firecrackers. Spring wind is sending the warmth to the Tusu drink." Hands in hands, they circled and chanted loudly.

    Little Ginkgo turned his head and asked, "Uncle Tony, did the firecracker in the ancient poem refer to the firecracker today?"

Hear the story of bamboo

History of Firecracker

Tony picked up a firecracker and showed Little Ginkgo the structure. He smiled and said, "No, they are not the same thing."

The name "firecracker" comes from the fact of burning the bamboo to let it burst. It is said that since the Northern and Southern Dynasties, on the lunar New Year's Day, the first thing every household did when opening the door was to burn bamboos. They hoped to use the crackling sound of bamboos to scare away the monster "Nian" and other evil spirits, so as to bless their families. Later, people invented gunpowder, made and burned firecrackers, which means getting rid of the old and welcoming the new. The sound of burst bamboo presages safety. This folk custom has a history of one thousand years or so.

森林的故事·竹子篇
The Story of the Forest · Bamboo Chapter

Little Ginkgo continued to ask, "Why is there a crackling sound when the bamboo culm is being burned?"

"The inside of bamboo culm is airtight as bamboo joints are distributed on it. What's more, green bamboo culm is rich in water. When burned, temperature rises, so that water evaporates. Continuous heating makes water vapor continue to expand, and the pressure inside the bamboo culm continues to increase. Finally, the bamboo cavity bursts, making a loud noise." explained Tony.

"Oh, I see! Only when green bamboo is burned, there will be a popping sound. If the bamboo is dry, it will not make a sound." Little Ginkgo was suddenly enlightened.

 Knowledge Point

History of Firecracker: Firecracker has a history of more than 2000 years in China. Before the invention of gunpowder and paper, Chinese people in ancient times burned the bamboo, making it burst to make a sound to expel the plague. The burst bamboo makes a "crackling" noise, so that it is called firecracker.

Chapter 6 Symbol of China
Bamboo and Giant Panda

森林的故事・竹子篇
The Story of the Forest · Bamboo Chapter

A Bamboo Eater

Since childhood, from storybooks to sculptures in the zoo, the national treasure—giant panda has left a deep impression on Little Ginkgo. They have big dark eye circles, long black scarves, chubby bodies, big round faces, and fat front paws holding green bamboo branches.

One day, as soon as he walked into the panda hall in the zoo, the first thing Little Ginkgo saw was a giant panda. It was sitting upright and cross-legged, with its front paws grasping the fresh bamboo and quickly stuffing it into its mouth, gnawing on it like eating sugar cane, making the sound of clicks, with its

Hear the story of bamboo

big cheeks puffing out, cute and naive.

Little Ginkgo turned his head and asked Uncle Tony, "Why do pandas like eating bamboos very much?"

"Is it because bamboos grow fast, and so fast into a big forest that it is convenient for pandas to eat?"

Tony was admiring the cute and naive giant panda. Before he could answer it, Little Ginkgo asked and answered himself.

"Correct, but not all correct." Tony's answer could always arouse Little Ginkgo's interest.

"Do you know that giant pandas, known as 'living fossils', have lived on the earth for 8 million years, and their ancestors were meat eaters or carnivores?"

森林的故事 · 竹子篇
The Story of the Forest · Bamboo Chapter

"Then how did they become bamboo-eaters later?"

"It was about 18000 years ago when the earth experienced dramatic changes in the natural environment, such as large-scale glaciers, which caused a sharp decline in forests and the subsequent extinction of many plants and animals. Lacking sufficient edible animals, pandas have survived all these years by gradually changing their diet."

"It turns out that bamboo is the life-saving food for giant pandas! It's no wonder that they have stuck to each other till now." Little Ginkgo found the answer.

Tony continued, "There are few animals that take bamboo as their staple food, so giant pandas have few competitors who eat the same food, which makes them have plenty of food. But they will occasionally seize the opportunity to catch a bamboo rat or poke a bird's nest to restore the carnivorous nature of their ancestors, and enjoy a little meat for themselves."

"Then the giant panda seems to be a 'fake monk' (a monk is supposed to be a vegetarian in China)!" Little Ginkgo teased.

Tony was amused by Little Ginkgo's humor.

Knowledge Point

Giant Panda and Bamboo: The giant panda, one of the oldest creatures in China, is known as a paleontological 'living fossil'. It lives mainly in the high-altitude areas of China and feeds on bamboos. The foreign gardening circle has identified bamboo and giant panda as the symbol of Chinese gardens.

A Strong Stomach

Uncle Tony and Little Ginkgo watched the giant panda's every move with interest. When the naughty giant panda was full, it excreted a big dung ball on the spot. Little Ginkgo unconsciously covered his nose with his hand.

Seeing Little Ginkgo's move, Tony joked, "The national treasure is special. Their feces are scented."

Little Ginkgo looked blank.

Tony explained with a smile, "Think about it, pandas have short intestines, the bamboo stays inside for a short time and is excreted before it is completely digested. If there is no fermentation, their feces will have the fragrance of bamboo."

Little Ginkgo tried his best to observe the feces, whose surface was laterally inserted. He vaguely saw slub fibers of different lengths. He worried whether bamboo scraps would scratch the digestive tract of the giant panda.

Little Ginkgo turned his head and found that Tony was still watching intently, so he didn't say a word. He saw that the giant panda bit the bamboo into small pieces, stuffed it into its mouth, and then chewed it together without spitting out scum.

Little Ginkgo couldn't resist to ask Tony. The latter also seemed to have guessed what he meant and said without haste, "Having eaten bamboo for a long time, giant pandas have evolved big and wide molars in their mouths and have strong masticatory muscles that help them effectively grind the rough bamboo fibers. Their digestive system has also become stronger. The thickness of their gastrointestinal mucosal muscles is about 8 times that of humans, which can help

Hear the story of bamboo

them better digest bamboo. However, they still can't fully digest fiber and lignin in bamboo."

"Therefore, bamboo leaves and fibers can be seen clearly in their feces. Will the sharp bamboo branches hurt giant pandas?" asked Little Ginkgo eagerly.

"The giant panda's digestive tract has well-developed mucus glands that secrete a lot of mucus. When the giant panda is feeding, the mucus wraps the bamboo to add lubrication and prevent damage to the digestive tract. During defecation, the mucus is wrapped around the feces, reducing the mechanical stimulation of crude fibers to protect giant pandas well." replied Tony.

Little Ginkgo, who likes to think, continued to ask, "If the bamboo is not fully digested, then the nutrients cannot be fully absorbed. But why are pandas still chubby?"

Tony said, "That's because our national treasures make great efforts to feed themselves. Compared to 'pure-blood' herbivores, the giant panda spends more than ten hours a day eating 20 kilograms of bamboo to replenish its body's energy, even though its utilization of plant nutrients is less than 20%. They spend rest of their time defecating about 40 times a day. It is also very hard."

After hearing this, Little Ginkgo opened his mouth wide in surprise and silently sighed, "It's not easy for national treasures to survive, we should all treasure them!"

Knowledge Point

Chemical Composition of Bamboo Stems: Bamboo stems are mainly composed of cellulose, hemicellulose and lignin. Generally, whole bamboo is composed of 50%—70% cellulose, 30% hemicellulose and 20%—25% lignin. In addition, it also contains a certain amount of protein, amino acids, lipids and so on. There are certain differences in the chemical composition of bamboo among various genera and species.

森林的故事·竹子篇
The Story of the Forest · Bamboo Chapter

Six Fingers

As soon as the weekend arrived, Little Ginkgo took Uncle Tony to the panda hall in the zoo again, lying on the fence to observe the giant pandas intently.

The giant panda in front of him was holding the bamboo and nibbling on it with relish. It grabbed the bamboo culm, along with the bamboo leaves, quickly stuffed it into its mouth, and then chewed it, making the sound of sizzle. Seeing the chubby giant panda deftly grasp and eat bamboo, Little Ginkgo couldn't help making a complaint, "The giant panda is a standard eater."

Suddenly, Little Ginkgo made a new discovery—the giant panda had six fingers. One, two, three, four, five, six! It really had six fingers! Little Ginkgo excitedly announced his new discovery to Tony, "Uncle Tony, the giant panda is the 'six-fingered bamboo demon'; it has six fingers. You see, there is a finger under the bamboo, and there are five fingers above it. It has one more finger than our human hands!"

Tony was amused by Little Ginkgo's expression and action and said approvingly, "That's right. The finger holding the bamboo is specially formed from a wrist bone, which is scientifically called 'radial cage bone'. Because the giant panda's claws are not as flexible as our fingers and in order to hold the bamboo, this wrist bone has become a new 'thumb' over time."

"Oh, I see." Little Ginkgo said with understanding.

"But this thumb is different from a human thumb. It is a carpal bone. There is no finger bone or joint in it, so it will not bend. It's just a 'pseudo-six finger'. Its function is to help the giant panda hold the bamboo steadily, and it is for sur-

Hear the story of bamboo

vival that 'the sixth finger' was evolved."

Little Ginkgo nodded in conviction and sighed, "Giant pandas are really smart! In order to be the best 'eater', they have continued to evolve, formed new traits, generated a new finger, and developed new skills in eating bamboo. This is also an example of 'survival of the fittest'!"

森林的故事·竹子篇
The Story of the Forest · Bamboo Chapter

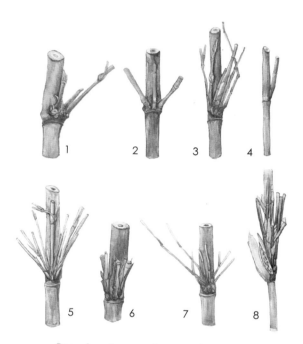

Some bamboo species eaten by giant pandas
1. *Chimonobambusa szechuanensis*　2. *Qiongzhuea rigidula*
3. *Yushania brevipaniculata*　4. *Indocalamus longiauritus*
5. *Fargesia pauciflora*　6. *Bashania fargesii*
7. *Fargesia denudata*　8. *Bashania fangiana*

 Knowledge Point

Edible Bamboo Species for Giant Pandas: There are more than 60 species of edible bamboos for giant pandas, mainly including *Bashania fangiana, Chimonobambusa szechuanensis, Qiongzhuea rigidula, Qiongzhuea tumidinoda, Qiongzhuea macrophylla, Indocalamus thessellatus, Fargesia pauciflora, Yushania brevipaniculata, Bashania spanostachya, Bashania fargesii, Fargesia scabrida, Fargesia denudata*, etc.

Bamboo Flowering and Famine

Chengdu Research Base of Giant Panda Breeding is the main research base of China for the relocation and protection of endangered wildlife such as giant pandas. While visiting the base, Little Ginkgo heard beautiful music being transmitted from the roadside stone-like speakers.

"The bamboo is blooming, hey, Mimi is lying in her mother's arms and counting stars. Oh, the star, the star, you are so beautiful! But where is my breakfast for tomorrow?..." Listening to the song, Little Ginkgo said happily, "I can sing it, I can sing it too." Then he couldn't help humming along.

Uncle Tony turned his head and asked Little Ginkgo, "Do you know the story behind the song '*Panda Mimi*'?"

Little Ginkgo shook his head and said, "I only know that Mimi is the name of the giant panda." Then he looked expectantly at Tony.

Tony pondered for a while and said, "This is a song that was popular in the 1980s. It was created to save the endangered national treasure, the giant panda."

"The panda was going to be extinct?" asked Little Ginkgo in surprise.

"At that time, the arrow bamboo forest in the north Min Mountains of Sichuan Province was blooming on a large scale, which made giant pandas living in the habitat face a severe food crisis, and then it was rare to see pandas there. To this end, the country has also launched a fundraising campaign for giant pandas."

"What is the connection between bamboo flowering and food crisis? Why does the flowering of bamboo cause pandas to run out of food?" asked Little Ginkgo suspiciously.

Hear the story of bamboo

森林的故事・竹子篇
The Story of the Forest · Bamboo Chapter

"Because the overall flowering of the bamboo usually causes the bamboo forest to wither and perish. When the bamboo forest is gone, pandas are left without food."

"After blooming, fruiting, and seeds sprouting, there should be the growth of new bamboos, how can they die?" Little Ginkgo was puzzled.

"Unlike ordinary plants, bamboo is a flowering plant and usually blooms once in a lifetime. In the early stage of growth, the vegetative growth of bamboo is dominant. Then, the growth advantage gradually turns to reproductive growth until it finally blooms and bears fruit. When they bloom and bear fruit, they need to consume a lot of organic nutrients stored in roots, stems and leaves. When nutrients in these vegetative organs are consumed, bamboos will die one after another." Tony explained patiently and carefully.

"It turns out that the flowering of bamboo has turned into the end of bamboo growth, which has led to the famine of the giant panda." Little Ginkgo suddenly realized.

"Therefore, we must also do a good job in the protection of bamboo resources and the sustainable development of bamboo management!" Tony patted Little Ginkgo, who nodded firmly.

 Knowledge Point

Flowering Habits of Plants: The flowering habits of plants can fall into two categories: one is once-flowering plants, such as rice, wheat, bamboo, etc.; the other is multiple-flowering plants, such as apples, pears, etc.

森林的故事·竹子篇
The Story of the Forest · Bamboo Chapter

Protecting Bamboos to Create a Better Future

It has been two weeks since Uncle Tony went out for investigation. Little Ginkgo missed him very much.

Coming home from school that day, Little Ginkgo looked up and saw a person sitting in front of the computer in the study, knowing that it was Uncle Tony who came back.

Little Ginkgo flew over, and suddenly sneaked between Tony and the desk like a loach. He sat on Tony's leg quickly and put his arms around his uncle's neck. He was so curious about the latter's scientific expedition, and asked hastily, "Uncle Tony, Uncle Tony, have you seen the wild giant panda?"

Looking at Little Ginkgo's eager expression, Tony pointed to the camera on the table and said with a smile, "How about coming with me to organize the photos of the expedition together?"

"Great! In this way, I can also follow the mysterious scientific expedition of giant pandas and bamboos through pictures!" Little Ginkgo replied excitedly.

Little Ginkgo methodically assisted Tony to import all photos from the camera into the computer and classify them according to time and place. Browsing pictures of bamboos, Tony said, "These are the favorite varieties of food for the national treasure, giant pandas."

"I know. There are more than 600 varieties of bamboo recorded in China, about 50 to 60 of which can be food of giant pandas but only over 20 of which

Hear the story of bamboo

are their favorite. Our national treasure is a picky eater."

Tony was amused by Little Ginkgo's metaphor. At this time, the picture of the giant panda lying on the branch displayed on the screen deeply attracted Little Ginkgo. The tree seemed to be more than ten meters high, and the location of the giant panda was at least seven or eight meters above the ground, with dense bamboo bushes underneath. Looking at the giant panda's chubby body and the top of the branch, Little Ginkgo really felt anxious for the national treasure, worrying that it would break the branch and fall off at any time.

Seeing Little Ginkgo's nervous expression with his wide eyes, Tony comforted him and said, "The giant panda is a master at climbing trees. The giant panda climbed up the tree after a full meal and was enjoying the good sunbathing. It was photographed in the deep forest of Sanguanmiao reserve in Foping, Qinling Mountains."

"Sanguanmiao, is it the place where it was reported a few days ago that there were five giant pandas holding a 'contest for marriage'?"

"Yes." Tony nodded and said approvingly, "It seems that our Little Ginkgo has collected a lot of panda 'intelligence'! What else have you mastered?"

"I also know that the threat level of giant pandas has been downgraded from 'endangered' to 'vulnerable'." said Little Ginkgo proudly.

"The drop in the protection level of giant pandas reflects our country's major achievements in biodiversity conservation and ecological restoration. The research project on the investigation and sustainable development of bamboo germplasm resources in the wild habitat of giant pandas this time is to better carry out the sustainable management of bamboo and better protect our national treasure."

"Yeah, I know. We must manage bamboo well and protect the national treasure."

"Yes. Only by strengthening ecological construction and doing a good job in the sustainable development of forest resources can giant pandas have a better home, a greener food, and a better future."

Tony put his palm on Little Ginkgo's shoulder and pressed it hard, as if he was passing on such an important responsibility to the younger generation.

 Knowledge Point

Sustainable Forest Management: Sustainable forest management refers to maintaining the health and vitality of forest ecosystem, preserving biodiversity and its ecological process through the scientific management and rational management of real and potential forest ecosystem, so as to meet the needs of forest products and their environmental service functions in the process of social and economic development, and ensure and promote the sustainable and coordinated development of population, resources, environment and social economy.

References

易同培. 四川竹类植物志 [M]. 北京：中国林业出版社，1997.

张齐生. 中国竹材工业化利用 [M]. 北京：中国林业出版社，1995.

江泽慧. 世界竹藤 [M]. 北京：中国林业出版社，2008.

康喜信，胡永红（上海植物园），等. 上海竹种图志 [M]. 上海：上海交通大学出版社，2011.

周芳纯. 竹林培育学 [M]. 北京：中国林业出版社，1998.

中国科学院中国植物志编辑委员会. 中国植物志 [M]. 北京：科学出版社，1991.

温太辉. 中国竹类彩色图鉴 [M]. 台北：淑馨出版社，1993.

周芳纯. 竹林培育学 [M]. 北京：中国林业出版社，1998.

彭镇华，江泽慧. 绿竹神气 [M]. 北京：中国林业出版社，2006.

李承彪. 大熊猫主食竹研究 [M]. 贵阳：贵州科技出版社，1997.

易同培，史军义，等. 中国竹类图志 [M]. 北京：科学出版社，2008.

周芳纯，胡德玉. 中国竹诗词选集 [M]. 南京：江苏古籍出版社，2001.

邹惠渝. 邵武竹类 [M]. 上海：上海科学技术文献出版社，1989.

陈守良，贾良智. 中国竹谱 [M]. 北京：科学出版社，1988.

史正军，杨静，杨海艳. 大型丛生竹材应用基础性能研究 [M]. 北京：科学出版社，2018.

王三毛. 古代竹文化研究 [M]. 北京：北京联合出版公司，2017.

Epilogue

One afternoon in the spring of 2018, the idea occurred to us that we should create a popular science book about bamboo after discussing it with Mr. Cao and Miss Zhou in the office for the first time. I am greatly delighted to hold out the manuscript of the book after four years of data collection, learning, reflection and research, summary analysis, concise induction, conceptual creation, writing, illustration and revision.

It is a book of popular science which integrates traditional culture and scientific knowledge, paper media and modern media, and gather the wisdom and talents of experts, scholars, teachers and students. The creation of this book has won the full guidance and strong support of Mr. Cao Fuliang. During preparation, Miss Zhou Jilin has done a lot of work in system audit, planning and coordination. Mr. Ding Yulong, Miss Zhang Chunxia and Director Yang Jian in China Giant Panda Protection and Research Center have reviewed scientific knowledge points. Miss Liu Dongbing and Mr. Zhang Wujun, etc. have participated in the editing of some texts. Bao Xinyue, Fu Junmin,

He Ling and Lu Tong, etc. have participated in the collection and compilation of materials. Wei Xin, Jia Wenting, Geng Zhirong, Zhao Yazhou and Song Yuxuan have taken responsibility for the creation and guidance of scenario maps and scientific maps. Cai Jiaqin, Li Yue, Zhang Jinyu, Liang Jiayu, Wu Xuan, Shu Hua have participated in the drawing of scene maps and scientific maps. The Modern Analysis and Test Center of NJFU has provided exquisite electron microscope pictures. Here, our sincere thanks will be given to all the teachers and students who have made contributions to the publication of this book.

Limited to the editor's knowledge and ability, there may be deficiencies in the book, so readers are kindly requested to comment and give suggestions.

November, 2022